THERMOSTABLE
PROTEINS

Structural Stability and Design

Srikanta Sen
Lennart Nilsson

CRC Press
Taylor & Francis Group
Boca Raton London New York

CRC Press is an imprint of the
Taylor & Francis Group, an **informa** business

CRC Press
Taylor & Francis Group
6000 Broken Sound Parkway NW, Suite 300
Boca Raton, FL 33487-2742

First issued in paperback 2017

© 2012 by Taylor & Francis Group, LLC
CRC Press is an imprint of Taylor & Francis Group, an Informa business

No claim to original U.S. Government works

Version Date: 20110818

ISBN-13: 978-1-4398-3913-3 (hbk)
ISBN-13: 978-1-138-11482-1 (pbk)

Library of Congress Cataloging-in-Publication Data

Thermostable proteins : structural stability and design / [edited by] Srikanta Sen,
  Lennart Nilsson.
       p. ; cm.
  Includes bibliographical references and index.
  ISBN 978-1-4398-3913-3 (hardcover : alk. paper)
  I. Sen, Srikanta. II. Nilsson, Lennart, 1952 Nov. 11-
  [DNLM: 1. Protein Stability. 2. Protein Conformation. 3. Thermodynamics. QU 55.9]

  LCclassification not assigned
  612'.01575--dc23   2011033944

Visit the Taylor & Francis Web site at
http://www.taylorandfrancis.com

and the CRC Press Web site at
http://www.crcpress.com

THERMOSTABLE
PROTEINS

Structural Stability and Design

Contents

Foreword

SETTING THE STAGE

Understanding the molecular basis of how thermostable and hyperthermostable proteins gain and maintain their stability and biological function at high temperatures remains an important scientific challenge. A more detailed knowledge of protein stability not only deepens our understanding of protein structure and function but also helps in obtaining insights into processes that drive protein folding, unfolding, and misfolding. Furthermore, as will be seen toward the end of this review collection, rules may be suggested for the design and engineering of thermostable proteins.

Numerous studies on the origins of protein thermostability have suggested a variety of factors that can provide favorable stabilizing contributions, including the hydrophobic effect, improved molecular packing, van der Waals interactions, networks of hydrogen bonds and ionic bonds, and optimized electrostatic interactions. It seems, however, that no dominating stabilization mechanism has evolved in proteins; rather, their thermostability results from a multitude of fine-tuned improvements of short- and long-range interactions. The role played by ionic interactions continues to be controversial, assuming that they are generally destabilizing, which may be more frequently true for proteins from mesophiles. Recent research on protein structures from hyperthermophiles has led to an important reevaluation of this assumption, demonstrating that these proteins are often characterized by an increased number of surface charges and ionic networks that may play dominant roles in enhancing extreme thermostability. To take into account the effects of temperature on the hydration of ionic bonds (these proteins are usually functional at the temperature of boiling water) has been of particular importance in this context. It has been shown that at those temperatures the desolvation penalty paid on the formation of a salt bridge decreases as a consequence of the lower dielectric constant of the solvent, thus leading to a favorable stabilizing contribution.

The basic question in research on protein thermostability is to grasp the physical principles, either thermodynamic or kinetic, that produce an increased thermal resistance at high temperatures. However, terms such as *thermostability*, *thermal stability*, *thermodynamic stability*, and *kinetic stability* have often led to misunderstandings or have been used in uncritical ways. Thus it seems worthwhile to point out the meaning of these stability measures. *Thermodynamic stability* is measured by the difference in free energy, ΔG, between folded and unfolded conformations under equilibrium conditions. It is assumed that the folding/unfolding transition obeys a two-state model and that the protein unfolds reversibly, which would allow the application of equilibrium thermodynamics. *Kinetic stability* depends on the rate of unfolding and is described by a rate constant that in turn depends on the energetic barrier separating folded and unfolded states. If a protein unfolds irreversibly it may follow more complex unfolding pathways, which would not allow analysis in terms of equilibrium thermodynamics. Irreversible folding/unfolding transitions are usually described by a melting temperature, T_m, and are (or should be) related to the term *thermal stability*. Unfortunately, the term *thermostability* is often used ambiguously. Its usage should always be linked to one of the stability measures defined above.

CURRENT RESEARCH FRONTIERS AND CHALLENGES

The role of fluctuations in protein thermostability has recently received considerable attention. The increased conformational entropy, which was detected in several cases by molecular dynamics simulations, is related to a broadened energy landscape allowing the thermophilic protein to sample more conformational states as the mesophilic counterpart, which in turn seems to be a consequence of the smaller unfolding heat capacity change, ΔC_p, found in abundance in proteins from hyperthermophiles. However, a sound interpretation of entropic or enthalpic fluctuations in terms of molecular structure is still lacking.

The information stated above leads directly to the existing controversy concerning the flexibility/mobility and rigidity of a protein structure and its stability: what is the cause for increased thermal tolerance structural flexibility or rather rigidity? Several experimental and simulation studies imply that thermal tolerance of a protein is not necessarily correlated with the suppression of internal fluctuations, usually the contrary is seen. Structural rigidity based on enthalpic content seems to offer only little thermodynamic advantage for enhanced protein stability. The concept of flexibility/mobility needs critical reconsideration and analysis, in which the vastly different timescales for the fluctuation modes of a protein molecule have to be taken into account.

If internal packing is expressed by a packing density of atoms or atomic groups, then hardly any differences would be found between protein structures from mesophiles, thermophiles, and hyperthermophiles. However, the literature is filled with reports claiming better packing in thermophilic and hyperthermophilic counterparts. Careful analysis of suitable physical quantities, like compressibility measurements, should help in resolving current controversies.

The mobility/rigidity dilemma needs to be considered from a further point of view: the role played by water molecules and networks in stability and the folding/unfolding transition of a protein. The role of exposed ion pairs, abundant in proteins from hyperthermophiles, in the protein–water interface is unclear. However, they should contribute to stability because of the reduced desolvation penalty at high temperature. A large number of mutational efforts have provided evidence that their existence is related to stability; however, ionic bonds may fluctuate between direct charge–charge interaction or water-mediated interaction, depending on hydration, as seen by molecular dynamics studies of solvated proteins. Recent studies point to the fact that the hydration layer of proteins from hyperthermophiles is characterized by an increased number of solvent–protein hydrogen bonds, which favor resistance to increased temperature. The often found highly charged surfaces of protein structures from hyperthermophiles would be in agreement with these findings.

A number of recent reports have claimed that charged networks in proteins from hyperthermophiles favor kinetic stability by giving a contribution to the kinetic barrier separating folded and unfolded states. It is assumed that due to surface clamping by the ionic interactions, the native, folded state of those proteins is kinetically trapped, which would result in slowing down of the unfolding reaction. It is still an open question as to how the increased thermal energy is dissipated by the structure into a larger number of accessible degrees of freedom, and in this way accommodating the disorder caused by increased thermal motion. Charged networks can exist in many different states. Could the increased configurational entropy play a role in this context?

THIS BOOK

The chapters presented in this book span a wide range of protein thermostability research. The basic structural, thermodynamic, and kinetic principles are covered and molecular strategies for the adaptation to high temperatures revealed by structure analysis are delineated. The roles of fluctuations, hydration, and internal packing are thoroughly discussed. In reading the text, the challenging nature of this research is obvious and needless to say, that advanced computing and simulation methods will play an increasingly important role in future thermostability research, especially in situations that cannot be tackled by experiments. One of the described examples is the promising computational approach based on rigidity theory that allows us to investigate and improve the thermal adaptation of a protein starting from a mesophilic counterpart. Enzymes with particular industrial importance, the subtilisin-like serine proteases, have been extensively studied by protein engineering. One chapter is devoted to the present state of knowledge concerning structure–function relations and the origin of their structural stability. Last, but not the least, computational and experimental approaches for the design of proteins with increased thermal stability based on sequences or three-dimensional structures are presented.

I feel quite sure that the science described in this book will stimulate both experimentalists and scientists involved in theoretical analysis and computing into a deeper understanding of the fascinating research field of protein thermostability and probably will evoke a timely reevaluation of existing controversies.

Rudolf Ladenstein
Karolinska Institute
Stockholm, Sweden

Contributors

Ryan M. Bannen
Department of Biochemistry and Great
    Lakes Bioenergy Research Center
University of Wisconsin–Madison
Madison, Wisconsin

Sohini Basu
Molecular Modeling Section
Biolab, Chembiotek
The Chattergee Group (TCG)
    Lifesciences Ltd.
Calcutta, India

Holger Gohlke
Institut für Pharmazeutische und
    Medizinische Chemie
Mathematisch-Naturwissenschaftliche
    Fakultät
Heinrich-Heine-Universität Düsseldorf
Düsseldorf, Germany

Doris L. Klein
Institut für Pharmazeutische und
    Medizinische Chemie
Mathematisch-Naturwissenschaftliche
    Fakultät
Heinrich-Heine-Universität Düsseldorf
Düsseldorf, Germany

Magnús M. Kristjánsson
Department of Biochemistry
Science Institute
University of Iceland
Reykjavík, Iceland

Simone Melchionna
Instituto Processi Chimice-Fisiei
Consiglio Nazionale delle Ricerche
Rome, Italy

Julie C. Mitchell
Departments of Biochemistry and
    Mathematics
Great Lakes Bioenergy Research Center
University of Wisconsin–Madison
Madison, Wisconsin

Atsushi Mukaiyama
Division of Biological Science
Graduate School of Science
Nagoya University
Nagoya, Japan

Lennart Nilsson
Department of Biosciences and
    Nutrition
Karolinska Institutet
Huddinge, Sweden

George N. Phillips, Jr.
Departments of Biochemistry and
    Mathematics
Great Lakes Bioenergy Research Center
University of Wisconsin–Madison
Madison, Wisconsin

Sebastian Radestock
Computational Structural Biology
    Group
Max-Planck-Institute of Biophysics
Frankfurt, Germany

Thomas J. Rutkoski
Departments of Biochemistry and
    Mathematics
Great Lakes Bioenergy Research Center
University of Wisconsin–Madison
Madison, Wisconsin

Srikanta Sen
Molecular Modeling Section
Biolab, Chembiotek
TCG Lifesciences Ltd.
Calcutta, India

Fabio Sterpone
Institut de Biologie Physico-Chimique
Laboratoire de Biochimie Téorique
Centre National de la Recherche
    Scientifique
Paris, France

Kazufumi Takano
Graduate School of Life and
    Environmental Sciences
Kyoto Prefectural University
Kyoto, Japan

1 Delineation of the Conformational Thermostability of Hyperthermophilic Proteins Based on Structural and Biophysical Analyses

Atsushi Mukaiyama and Kazufumi Takano

CONTENTS

1.1   INTRODUCTION

A variety of microorganisms grow in physically or geochemically severe conditions. Microorganisms are generally classified into four groups on the basis of the temperature at which they grow: psychrophiles, mesophiles, thermophiles, and hyperthermophiles. Hyperthermophiles, which grow optimally at temperatures greater than 80°C, are found in hot environments such as deep-sea vents, submarine hydrothermal areas, and continental solfataras.[1] In the late 1960s, the first hyperthermophiles, *Sulfolobus* sp., were discovered in a hot acidic spring in Yellowstone National Park.[2] Since then, more than 50 hyperthermophile species have been discovered and it has been reported that a hyperthermophile can survive even at 122°C.[3] Most hyperthermophiles are members of the Archaea (e.g., *Thermococcus*, *Pyrococcus*, and *Sulfolobus*) and some bacterial hyperthermophiles (*Thermotoga* and *Aquifex*) have been discovered.

Hyperthermophilic proteins generally exhibit greater stability to function at temperatures near the boiling point than the equivalent proteins in microorganisms growing at moderate temperatures. The ability of hyperthermophilic proteins (enzymes) to function at high temperature is of great interest for biotechnology, and some proteins from (hyper)thermophiles have provided major breakthroughs in molecular biology.[4–6] In spite of the importance of hyperthermophilic proteins, there is a paucity of studies in which the stability of hyperthermophilic proteins have been analyzed in detail.

Here we describe our recent studies of the conformational stability of hyperthermophilic proteins using biophysical and crystallographic techniques. First, we focus on the thermodynamic analysis of the effects of various factors on the stability of a hyperthermophilic protein. Thermodynamic stability in relation to molecular evolution is discussed. Next, we describe the structure-based analysis of the stability of hyperthermophilic proteins. Together, these results provide novel insights into the stabilization of proteins. On the basis of these results, we propose the significance of further studies of hyperthermophilic proteins.

1.2   THERMODYNAMICS OF THE STABILIZATION MECHANISM OF A HYPERTHERMOPHILIC PROTEIN

Examining the thermodynamic stability of proteins is useful and important for understanding the molecular basis of the stabilization mechanism of proteins. The Gibbs free energy change (ΔG) between the native and the denatured states upon unfolding is a reasonable measure for the evaluation of protein stability.[7–11] In the case of a protein with an unfolding reaction that is assumed to be a simple two-state process, ΔG is expressed by

$$\Delta G = -RT \ln K, \qquad K = [U]/[N] \tag{1.1}$$

where U is the unfolded state, N is the native state, R is the gas constant, T is absolute temperature (in Kelvin), and K represents the temperature and equilibrium constant between the N and U states. To obtain ΔG experimentally, it is necessary to perturb the equilibrium between the N and U states by changing the temperature, the pH, or the concentration of a denaturant such as urea or guanidine hydrochloride (GdnHCl), and to follow a shift in equilibrium by using biophysical techniques such as differential scanning calorimetry (DSC), fluorescence spectroscopy, or circular dichroism (CD) spectroscopy.[12] A denaturant-induced equilibrium unfolding transition is often investigated to obtain ΔG at a certain temperature. Assuming a two-state process, the unfolding transition curve is analyzed as

$$y = \{(b_n + a_n[D]) + (b_u + a_u[D])\exp[(\Delta G(H_2O) - m[D])/RT]\}/$$
$$\{1 + \exp[(\Delta G(H_2O) - m[D])/RT]\}$$

(1.2)

where y is the observed signal at a given concentration of denaturant, [D]; b_n is the signal for the native state and b_u is the signal for the unfolded state; a_n is the slope of the pretransition of the baseline and a_u is the slope of the posttransition of the baseline; $\Delta G(H_2O)$ is the value of ΔG in the absence of GdnHCl; and m is the slope of the linear correlation between ΔG and [D]. Earlier studies demonstrated that hyperthermophilic proteins have higher $\Delta G(H_2O)$ values than their mesophilic counterparts because of differences in the number of ionic interactions, the extent of hydrophobic surface burial, and various combinations of these factors.[13–15] In addition, the temperature dependence of ΔG gives further insight into protein stability and the contributions of enthalpy (ΔH) and entropy (ΔS) to ΔG upon unfolding. The temperature dependence of ΔG is expressed as

$$\Delta G(T) = \Delta H_m - TH_m/T_m + \Delta C_p[T - T_m - T\ln(T/T_m)]$$

(1.3)

where ΔH_m is the enthalpy of unfolding at the transition midpoint temperature (denaturation temperature, T_m) and ΔC_p is the difference in heat capacity between the native and the denatured states. There are three models proposed to explain how hyperthermophilic proteins can adapt to higher temperatures than the equivalent proteins from organisms growing in moderate temperatures: (1) up-shifting, (2) right-shifting, and (3) flattening of the stability curve.[16–22]

ΔG upon unfolding can be determined indirectly from kinetic unfolding/refolding experiments in which an equilibrium is shifted by rapidly changing the temperature, pH, or denaturant concentration and the time course of unfolding/refolding reactions is monitored by the techniques described above.[23] In the case of a protein showing a simple two-state transition with no transient intermediate, the observed unfolding/refolding curves are analyzed by using a single exponential, yielding a single apparent rate constant k_{app}, where $k_{app} = k_{unf} + k_{ref}$, and k_{unf} and k_{ref} represent the unfolding and refolding rate constants. In the kinetics experiments using the denaturant concentration jump method, the logarithm of k_{app} of the unfolding and refolding depends linearly on the final concentration of the denaturant. The dependence of $\ln k_{app}$ on the concentration of the denaturant, called a chevron plot, is generally analyzed using the equation

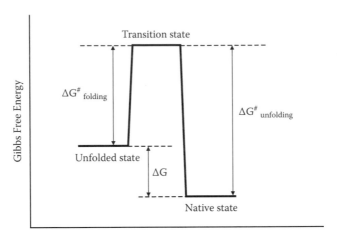

FIGURE 1.1   A simplified reaction diagram for two-state protein folding based on the transition state theory. $\Delta G^{\#}_{\text{folding}}$ and $\Delta G^{\#}_{\text{unfolding}}$ denote the activation energy between the unfolded and transition states and between the respective native and transition states.

$$\ln k_{\text{app}} = \ln \{k_{\text{ref}}(H_2O)\exp(-m_{\text{ref}}[D]) + k_{\text{unf}}(H_2O)\exp(+m_{\text{unf}}[D])\} \qquad (1.4)$$

where $k_{\text{unf}}(H_2O)$ and $k_{\text{ref}}(H_2O)$ represent the unfolding and refolding rate constants in the absence of the denaturant, m_{unf} and m_{ref} are their dependence on the concentration of the denaturant, and [D] is the final concentration of denaturant. Again, if there is no kinetic intermediate during either the unfolding or the refolding process, ΔG is obtained by using the rate constants of the unfolding and the refolding reactions: $\Delta G = RT\ln K$, where $K = k_{\text{ref}}/k_{\text{unf}}$.

From the evaluation of protein stability in terms of both equilibrium and kinetic aspects, we can discuss how hyperthermophilic proteins are more kinetically stabilized compared to the equivalent proteins in organisms growing at moderate temperature.[24] The increased stability of hyperthermophilic proteins should be caused by the slower unfolding rate (the larger the $\Delta G^{\#}_{\text{unfolding}}$ value), the faster the refolding rate (the smaller the $\Delta G^{\#}_{\text{folding}}$ value), or both (Figure 1.1). Such a bidirectional analysis is expected to highlight novel energetic features of the conformational stability of hyperthermophilic proteins.

1.2.1   RIBONUCLEASE HII FROM *THERMOCOCCUS KODAKARAENSIS*

The conformational stability of ribonuclease HII from the hyperthermophilic archaeon *Thermococcus kodakaraensis* (Tk-RNase HII) has been studied extensively by the authors.[25–29] Ribonuclease H (RNase H) hydrolyzes the phosphodiester bonds of RNA hybridized to DNA at the PO-3′ bond.[30] Tk-RNase HII is a monomeric protein consisting of 228 amino acid residues, and the crystal structures of this protein and its several mutants have been determined.[31–33] We first investigated the GdnHCl-induced unfolding transition of Tk-RNase HII to determine the ΔG value of this protein. The reaction was monitored by measuring molecular ellipticity at 220 nm, which reflects the contents of the secondary structure of the protein. To verify the reversibility of the reaction, the unfolding transition curve obtained by mixing a buffered solution of GdnHCl with a solution containing native Tk-RNase HII was compared to the refolding transition curve obtained by mixing the buffered solution of GdnHCl with a solution containing denatured Tk-RNase HII and a large amount of GdnHCl. Subsequently the GdnHCl-induced unfolding transition of Tk-RNase HII was found to be highly reversible at all temperatures examined. The equilibrium unfolding transition induced by GdnHCl was assumed to be a two-state process, with a reaction at 50°C attaining equilibrium in 2 weeks (Figure 1.2a). $\Delta G(H_2O)$ and m values at 50°C were found to be 43.6 kJ/mol and 23.6 kJ/mol/M, respectively, which is indicative of the greater stability of this protein compared with most mesophilic proteins.[34] The unfolding transition curve did not match the refolding transition curve even after 30 days at 20°C (Figure 1.2b). The discrepancy between these transition curves suggests that Tk-RNase HII exhibits slower unfolding, refolding, or both. Finally, it took about 2 months to reach equilibrium at 20°C.

Next, we examined the temperature dependence of $\Delta G(H_2O)$ (stability curve) for Tk-RNase HII and compared it with those for the mesophilic bacterium *Escherichia coli* (Ec-RNase HI; Figure 1.3) and the thermophilic bacterium *Thermus thermophilus* (Tt-RNase HI; Figure 1.3).[25,35] Tt-RNase HI increases the stability by shifting the stability curve up and flattening it compared to that of Ec-RNase HI. The stability curve for Tk-RNase HII has a maximum at approximately 40°C, which is higher than that of both Ec-RNase HI and Tt-RNase HI. The ΔC_p value of Tk-RNase HII (14.5 kJ/mol/K) is larger than that of Tt-RNase HI. Therefore Tk-RNase HII adapts to higher temperature by shifting the stability curve up and to the right. The results suggest that the mechanisms used by Tk-RNase HII and Tt-RNase HI to adapt to higher temperatures are different.

The heat-induced unfolding transition of Tk-RNase HII was examined by DSC,[25] which measures heat change directly during heating or cooling and is a powerful technique for understanding protein thermal stability. DSC curves obtained at scan rates of 30°C/hr and 60°C/hr are shown in Figure 1.4. The reversibility of this reaction was confirmed by reheating experiments. The repeated thermal scans were superimposable, indicating that the heat-induced unfolding of Tk-RNase HII is highly reversible. Such high reversibility of GdnHCl- and heat-induced unfolding shows clearly that Tk-RNase HII is suitable for analyzing the thermodynamic stability of proteins. As shown in Figure 1.4, Tk-RNase HII retained the native structure at greater than 80°C, but the denaturation temperature in the DSC curve was found to be dependent on the scan rate. The denaturation temperature at a scan rate of 90°C/hr was 89.2°C, whereas it was 87.2°C at a scan rate of 5°C/hr. This observation means that the heat-induced unfolding reaction does not attain equilibrium under the conditions examined. Similar observations were reported for pyrrolidone carboxyl peptidase from *Pyrococcus furiosus*,[36] whereas the heat-induced unfolding reaction can attain equilibrium for most mesophilic proteins at a scan rate of 60°C/hr. This slower unfolding appears to be a characteristic of the stabilization mechanism of Tk-RNase HII.

To delineate the kinetic aspects of Tk-RNase HII, we examined the kinetic unfolding and refolding induced by the GdnHCl concentration jump at 50°C.[25]

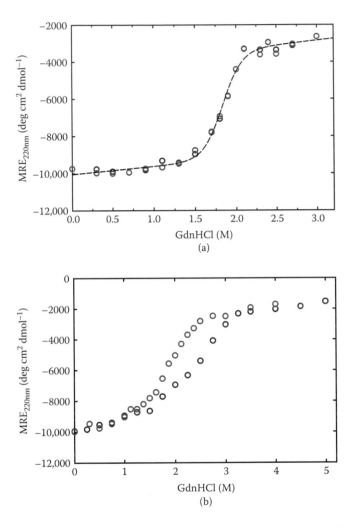

FIGURE 1.2 GdnHCl-induced equilibrium unfolding (blue circles) and refolding (red circles) curves of Tk-RNase HII at pH 9.0 for (a) 50°C for 2 weeks and (b) 20°C for 1 month. The reaction was followed by measuring CD at 220 nm. The broken line in (a) is the theoretical curve based on Equation (1.2). For the unfolding curve, Tk-RNase HII was incubated with different concentrations of GdnHCl. For the refolding curve, the protein, which was unfolded completely in 4 M GdnHCl, was diluted with buffer and the diluted protein solution was incubated. **(See color insert.)**

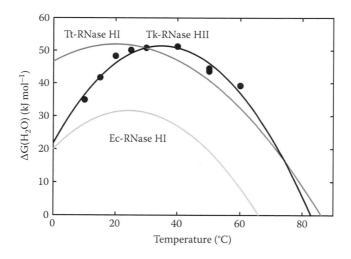

FIGURE 1.3   Stability curves for Tk-RNase HII (orange), Ec-RNase HI (cyan), and Tt-RNase HI (red) obtained from GdnHCl-induced equilibrium unfolding experiments. Filled circles are experimental data for Tk-RNase HII, and all lines represent the theoretical curves based on the Gibbs–Helmholtz equation (Equation 1.3). **(See color insert.)**

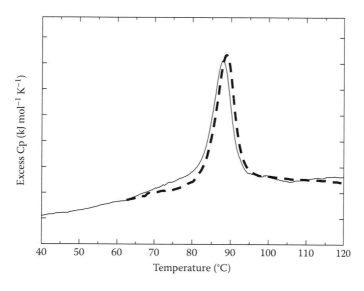

FIGURE 1.4   DSC curves for Tk-RNase HII at scan rates of 30°C/hr (unbroken line) and 60°C/hr (dotted line).

Figure 1.5A shows representative kinetic unfolding curves of Tk-RNase HII monitored by measuring the molecular ellipticity at 220 nm. The kinetic traces of unfolding and refolding reactions were all fit to a single exponential, which yielded apparent rate constants (k_{app}) for the reaction. The natural logarithm of k_{app} as a function of the final concentration of GdnHCl is shown in Figure 1.5B. The $k_{unf}(H_2O)$ and $k_{ref}(H_2O)$ at 50°C are 5.0×10^{-8}/s and 7.8×10^{-1}/s, respectively, and m_{unf} and m_{ref} at 50°C are 2.8/M/s (7.5 kJ/mol/M) and 5.5/M/s (14.8 kJ/mol/M), respectively. The $\Delta G(H_2O)$ value of 44.5 kJ/mol obtained from these rate constants

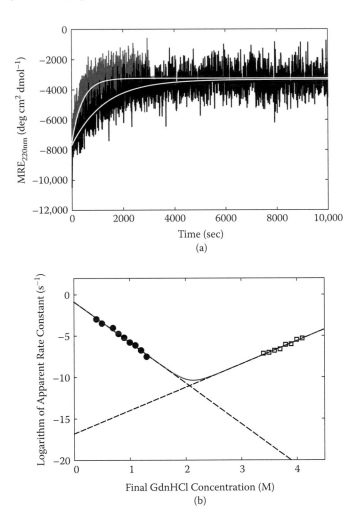

FIGURE 1.5 Kinetic experiments for Tk-RNase HII at 50°C and pH 9.0. (a) Kinetic unfolding curves for Tk-RNase HII monitored by a change in CD at 220 nm. The reaction was initiated by GdnHCl concentration jumps to 3.9 M (red line) and 3.4 M (black line). **(See color insert.)** (b) GdnHCl concentration dependence of the apparent rate constants (k_{app}) of unfolding (open squares) and refolding (filled circles) reactions. The unbroken line represents the theoretical curve using Equation (1.4).

using a two-state model at 50°C is similar to the value of 43.6 kJ/mol that was obtained from the equilibrium unfolding experiment. This suggests two-state folding/unfolding of Tk-RNase HII.

The unfolding and refolding kinetics of Tt-RNase HI and Ec-RNase HI at 25°C have been well characterized.[37–39] Thus, we examined the kinetic unfolding and refolding of Tk-RNase HII at 25°C and compared them with those of Tt-RNase HI and Ec-RNase HI. The unfolding rate constant for Tk-RNase HII at 25°C (6.0 × 10^{-10}/s) is much smaller than those for Ec-RNase HI (1.1 × 10^{-5}/s) and Tt-RNase HI (4.0 × 10^{-6}/s). In contrast, a large difference was not observed among these proteins in the refolding rate constant for the formation of the native state. Tk-RNase HII has equilibrium stability similar to that of Tt-RNase HI, whereas the unfolding rate constant for Tk-RNase HII is much smaller than that for Tt-RNase HI. This suggests that the greater stability of Tk-RNase HII originates from slower unfolding.

1.2.2   FACTORS AFFECTING THE THERMODYNAMIC STABILITY OF HYPERTHERMOPHILIC PROTEINS

As shown above, our study showed that the slower unfolding is one of the major causes of the greater stability of Tk-RNase HII. This trend is not restricted to Tk-RNase HII; there are some reports of hyperthermophilic proteins exhibiting slower unfolding. For example, the cold shock protein from the hyperthermophilic bacterium *Thermotoga maritima* (Tm-Csp) displayed a higher equilibrium stability at 20°C compared to those from the thermophilic bacterium *Bacillus caldolyticus* (Bc-Csp) and the mesophilic bacterium *Bacillus subtilis* (Bs-Csp).[40] The unfolding rate constant of Tm-Csp was slower than those from Bc-Csp and Bs-Csp, whereas the refolding rate constants were similar among these proteins. The slower unfolding of Tm-Csp compared to Bs-Csp was observed over a wider range of temperatures. The difference was found to be due to a difference in the activation entropy of unfolding.[41] In addition, it has been reported that the greater stability of ribosomal protein S16 from the hyperthermophilic bacterium *Aquifex aerolicus* compared to its mesophilic homologue originates from the slower unfolding.[42] Therefore slower unfolding appears to be a determinant of the greater stability of hyperthermophilic proteins. However, many reviews about the stability of hyperthermophilic proteins focus on the molecular mechanism in terms of the equilibrium aspect, not the kinetic aspect, because of the lack of experimental data for the unfolding/refolding kinetics of hyperthermophilic proteins.[10,11] Here, we describe factors contributing to the slower unfolding of hyperthermophilic proteins based on the results obtained for Tk-RNase HII.

1.2.2.1   Hydrophobic Interactions

The hydrophobic effect is known to be one of the most important factors for retention of the folded structure in proteins and its contribution to stability has been well characterized.[43–47] We used mutational analysis for Tk-RNase HII to examine the effect of hydrophobic interactions on the equilibrium and kinetic stability of this protein.[27] Mutations in which a large buried hydrophobic side chain is replaced by a smaller one (Lue/Ile to Ala) decreased the stability significantly by 8.9 to 22.0 kJ/

mol at 50°C. In kinetic experiments, these mutations affect mainly the unfolding speed, and the rate constants for these mutant proteins are from one to three orders of magnitude greater than that of the wild-type Tk-RNase HII. These mutations slow refolding, but the differences between the wild-type and the mutant proteins are less than one order of magnitude. These results demonstrate that the hydrophobic effect contributes to the slower unfolding of Tk-RNase HII. In other words, it is suggested that the hydrophobic interactions contributing to the slower unfolding are mainly disrupted before passing the unfolding transition state. This is the first report that presents a practical cause of the slow unfolding of hyperthermophilic proteins.

1.2.2.2   Proline Effect

It is known that proline residues have lower configuration entropies than any other amino acid residues.[48] Proline residues decrease the conformational entropy of the denatured state of proteins, leading to an increase in stability.[49] It has been reported that proline residues introduced into N-terminal α helices increase the thermostability of proteins.[50] Comparisons of sequences among homologous proteins exhibiting different levels of stability demonstrate that proline residues at the N-terminus of an α helix are frequently found in more stable proteins. Indeed, there are some proline residues located at the N-terminus of the α helix of Tk-RNase HIII. We used systematic mutational analysis of Tk-RNase HII to examine the proline effect.[28] Mutation of proline residues at the N-terminal α helices had only a minor effect on the equilibrium stability of Tk-RNase HII. In contrast, the unfolding rate for the mutant proteins was similar to that of the wild-type protein, indicating that proline residues contribute little to the slow unfolding of Tk-RNase HII.

1.2.2.3   Naturally Occurring Osmolytes

The maintenance of an appropriate turgor pressure in cells is achieved by the accumulation of naturally occurring osmolytes or those synthesized by cells. These osmolytes are originally small organic molecules that are known to act as protein stabilizers.[51,52] Osmolytes are ubiquitously present in plants, animals, and microorganisms, including hyperthermophiles. Trimethyl-amine-*N*-oxide (TMAO) is a naturally occurring osmolyte, and the equilibrium stability of Tk-RNase HII was found to be increased by the addition of 0.5 M TMAO.[29] In kinetic unfolding and refolding experiments using the GdnHCl concentration jump method, the presence of 0.5 M TMAO slowed the unfolding rate and increased the refolding rate to a similar extent, which resulted in increased equilibrium stability. Such an effect of osmolytes was observed in some proteins whose host organisms grow at moderate temperatures,[53,54] which indicates that the decrease in the unfolding rate constant induced by the presence of osmolytes is not intrinsic to hyperthermophilic proteins.

1.2.3   Implications for the Relationship between the Evolution and Stability of Hyperthermophilic Proteins

The significant difference in the unfolding speed and the similar equilibrium stability of Tk-RNase HII and Tt-RNase HI have provided the motivation for further investigation into the cause of the distinct stabilization mechanism. There are two major differences between the two proteins: the kingdom of life of their host organism and the type of RNase H. Tk-RNase HII is a type 2 RNase H and its host, *T. kodakaraensis*, belongs to the Archaea. Tt-RNase HII is a type 1 RNase H and its host, *T. thermophilus*, belongs to the Bacteria. Such differences might be due to a difference in the unfolding rate between Tk-RNase HII and Tt-RNase HI. To test this hypothesis, we examined the equilibrium stability and kinetic unfolding reaction of RNases HII from the hyperthermophilic bacteria *T. maritima* (Tm-RNase HII) and *A. aerolicus* (Aa-RNase HII) and RNase HI from the hyperthermophilic archaeon *Sulfolobus tokodaii* (Sto-RNase HI).[55] These proteins are all more stable than RNase HI from *E. coli*, and Sto-RNase HI was the most stable of the four hyperthermophilic proteins. Unfolding rate constants for these proteins at 25°C were widely distributed within a range of six orders of magnitude, as shown in Figure 1.6. The $k_{unf}(H_2O)$ value for Aa-RNase HI was comparable with that for RNase HI from *E. coli*, whereas Tk-RNase HII and Sto-RNase HI exhibited significantly slower

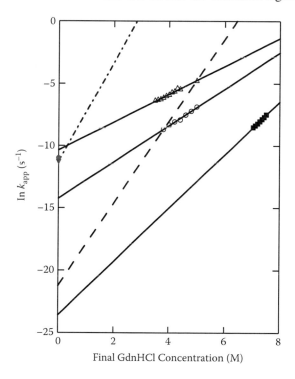

FIGURE 1.6 GdnHCl concentration dependence of $\ln k_{app}$ for unfolding of RNases H at 25°C. Open circles represent the data for Tm-RNase HII at pH 7.5; open triangles represent the data for Aa-RNase HII at pH 5.0; and filled squares represent the data for Sto-RNase HI at pH 3.0. The unbroken lines represent the linear fit. The long-dash line represents the data for Tk-RNase HII and the short-dash line represents the data for Ec-RNase HI.[25,37] The red circles and blue triangles represent the $k_{unf}(H_2O)$ value obtained from urea-induced unfolding experiments with Ec-RNase HI[38] and Tt-RNase HI,[39] respectively. **(See color insert.)**

unfolding rate constants. In other words, unfolding rate constants for thermophilic archaeal proteins are smaller than those for thermophilic bacterial proteins, implying that the unfolding speed of thermophilic proteins is determined by the evolutionary history of the host organism. Comparison of the crystal structures of these proteins suggests that a difference in unfolding rate constant is related to the amount of buried hydrophobic residues in the tertiary structure, which is consistent with the experimental result that hydrophobic interactions greatly affect the unfolding rate constant for Tk-RNase HII.

1.3 MOLECULAR STRATEGY OF ADAPTATION TO HIGH TEMPERATURE REVEALED BY STRUCTURAL ANALYSIS

Because of advances in molecular biology, the genomes of many hyperthermophiles have been sequenced. It has been reported that an increase in the G + C content was observed in nucleotide sequences of hyperthermophiles compared to those of mesophiles.[56,57] In recent years, crystal structures of increasing numbers of hyperthermophilic proteins have been determined, which suggests that various factors, including increased numbers of salt bridges, improved hydrogen bonding, favorable packing interactions, fewer cavities, and improved hydrophobic interactions, are involved in the stability of hyperthermophilic proteins.[58–62] Further crystallographic analysis might find as yet unknown strategies for the thermostability of proteins. Here we report studies of the molecular machinery of thermostability in subtilisin-like serine proteases and glycerol kinase from *T. kodakaraensis* as revealed by crystallographic analysis.

1.3.1 SUBTILISIN-LIKE SERINE PROTEASES FROM *THERMOCOCCUS KODAKARAENSIS*

Subtilisin-like serine protease hydrolyzes peptide bonds with broad substrate specificity and is widely distributed in various organisms, including members of the Bacteria, Archaea, and Eucaryota. This protein is synthesized in a precursor form consisting of a signal peptide that is involved in secretion into the external medium, a propeptide, and a mature domain. Bacterial subtilisins, such as subtilisin E and subtilisin BPN, whose folding and maturation process have been well studied,[63–73] are secreted in a pro form with the assistance of a signal peptide and are activated upon autoprocessing and degradation of the propeptide. For the mature domain to be functional, the propeptide must be degraded, because it binds to the mature domain tightly after autoprocessing and inhibits its activity. Subtilisin-like serine proteases are secretory proteins that need to adapt to an ambient temperature without intercellular stabilization factors such as osmolyte and crowding effects. Therefore it is expected that novel strategies for a stabilization mechanism are to be found in subtilisin-like serine protease from hyperthermophiles.

We report on two subtilisin-like serine proteases from *T. kodakaraensis* (Tk-subtilisin and Tk-SP), whose genome contains three genes that encode subtilisin-like serine proteases. These two proteins are functionally and structurally similar, but they appear to have different strategies for adaptation to a hot environment. We discuss the details below.

1.3.1.1   Tk-Subtilisin

Tk-subtilisin has been investigated extensively with regards to its function, folding, and structure.[74–79] Tk-subtilisin consists of a signal peptide (Met-24 to Ala-1), a propeptide (Gly1 to Leu69), and a mature domain (Gly70 to Gly398). The amino acid sequence of Tk-subtilisin is similar to those of bacterial subtilisins, except that the mature domain of Tk-subtilisin contains three major insertion sequences. It has been shown that mature Tk-subtilisin has the greatest enzymatic activity at 90°C with a half-life of approximately 9 hr. Considering that subtilisin E loses its enzymatic activity at 60°C with a half-life of 18 min, Tk-subtilisin was found to be very thermostable. Biochemical analysis demonstrated that the maturation process of Tk-subtilisin is essentially consistent with that of bacterial subtilisins, except for the role of Ca^{2+} and the propeptide for folding. In bacterial subtilisins, Ca^{2+} is required for stability, but not for activity or folding,[64,80–82] whereas Ca^{2+} is necessary for folding and activity in Tk-subtilisin.[75] Furthermore, crystallographic analysis has demonstrated a remarkable difference in structure between Tk-subtilisin and bacterial subtilisins.[76,77] Seven Ca^{2+}-binding sites were observed in Tk-subtilisin, but only two in the bacterial proteins (Figure 1.7). The increase of bound Ca^{2+} appears to be utilized for Tk-subtilisin in adapting to high temperature. Four calcium ions are formed at a loop region (Gly206 to Glu229) that consists mostly of the unique insertion sequence of this protein. A mutant protein devoid of this Ca^{2+}-binding loop was unable to refold to the native structure. In contrast, the mutant proteins lacking the Ca^{2+}-2 or Ca^{2+}-3 binding site could refold into native structures, albeit much more slowly than Pro-Tk-subtilisin (Pro-S324A).[78] Comparison of DSC curves for these three proteins showed that both mutant proteins lacking a Ca^{2+}-binding site were slightly more stable than Pro-Tk-subtilisin, suggesting that the unique Ca^{2+}-binding loop does not contribute greatly to the stability of this protein but is necessary for folding Tk-subtilisin.

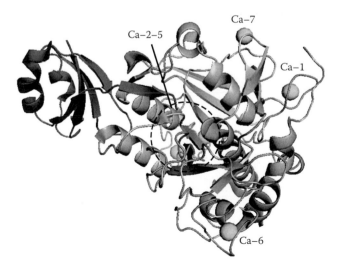

FIGURE 1.7   The crystal structure of mutant Tk-subtilisin mimicking its autoprocessed form. Propeptide and mature domain are represented in red and gray, respectively, and Ca^{2+} is shown in cyan. (**See color insert.**)

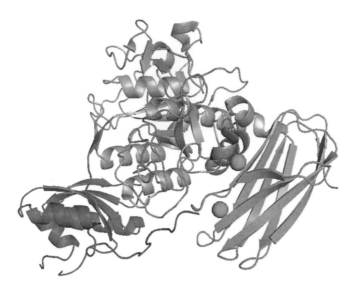

FIGURE 1.8 The crystal structure of the active site mutant of Pro-Tk-SP lacking a C-terminal propeptide. The N-terminal propeptide, mature domain, and extended C-terminal region are represented in red, gray, and green, respectively, and Ca^{2+} is shown in cyan. **(See color insert.)**

1.3.1.2 Tk-SP

Tk-SP is activated upon autoprocessing and degradation of N-terminal and C-terminal propeptides during the maturation process. Similar to Tk-subtilisin, Tk-SP exhibited a high level of tolerance of high temperatures with a half-life of 100 min at 100°C.[83] Unlike Tk-subtilisin and other bacterial subtilisins, Tk-SP requires neither Ca^{2+} nor propeptide for folding, and it has a long C-terminal extension. Without this extension, the amino acid sequences of Tk-SP and Tk-subtilisin are quite similar (41% identity). These results suggest that the stabilization mechanism of this protein is quite different from that of Tk-subtilisin. The crystal structure of Pro-Tk-SP lacking a C-terminal propeptide shows that the overall structure of this protein is superimposable on that of Tk-subtilisin, except that it contains no Ca^{2+}-binding site in the N-terminal propeptide and mature domain, and that the extended C-terminal region adopts a β jelly roll fold with two Ca^{2+}-binding sites (Figure 1.8).[84] Removal of this β jelly roll domain had only a minor effect on the folding and activity of Tk-SP, whereas a significant decrease in the T_m value of Tk-SP lacking the β jelly roll domain was observed in the presence of Ca^{2+}. These results indicate that Tk-SP adopts a unique strategy, attachment of a β jelly roll domain to the C-terminus, in order to adapt to hot environments.

1.3.2 GLYCEROL KINASE FROM *THERMOCOCCUS KODAKARAENSIS*

Glycerol kinase (GK) catalyzes the ATP-dependent phosphorylation of glycerol to produce glycerol 3-phosphate, which is important as a metabolic intermediate for glycolysis. GK from *T. kodakaraensis* (Tk-GK) shows a high level of stability against heat treatment and retains its enzymatic activity during incubation at 90° for 60 min.[85] Tk-GK forms a dimer in solution, unlike its mesophilic counterpart GK from *E. coli* (Ec-GK), which exists in a dimer–tetramer equilibrium.[86] Comparison of the crystal structures of Tk-GK and Ec-GK showed that Tk-GK has four additional ion pairs in the α16 helix.[87] Reciprocal mutations in the α16 helix were applied to Tk-GK and Ec-GK in order to disrupt or create the ion pairs. As a result, Ec-GK was stabilized by this mutation, whereas the mutations did not seriously affect the stability of Tk-GK. These results suggest that the difference in stability between Tk-GK and Ex-GK is not based on a simple mechanism, and that the ion pairs in the α16 helix contribute to the thermostability of Tk-GK in a cooperative manner.

1.4 FUTURE PERSPECTIVE

In this chapter we have described the results of our study on the stability of hyperthermophilic proteins. We found various novel features of hyperthermophilic proteins that are not easily accessible in proteins from organisms that grow at moderate temperatures. We found, through a detailed thermodynamic analysis of protein stability, that the higher stability of hyperthermophilic proteins with reversible unfolding is characterized by kinetic stability. Hydrophobic interactions are one of the factors that contribute greatly to kinetic stability. There are various factors that potentially affect the slower unfolding (e.g., the charge–charge interaction and protein–water interaction). Hence the contribution of these factors to the kinetic stability of hyperthermophilic proteins is an interesting subject for future study. Crystallographic analysis is a powerful method for characterizing the nature of the stability of hyperthermophilic proteins because it provides valuable information at the atomic level. Here, structural analysis of Tk-subtilisin and Tk-SP showed they are highly homologous, whereas the strategies they use to adapt to high temperatures are totally different. In addition, comparison of the crystal structures revealed a marked difference between Tk-GK and Ec-GK in the number of ion pairs in a helix. However, it has been shown by biochemical and biophysical analysis that such a difference does not explain all of the difference in stability between these proteins. These results indicate the importance of such integral analysis. We believe that hyperthermophilic proteins are a treasure trove of as yet unknown insights about the stabilization mechanism of proteins. Further studies of hyperthermophilic proteins will deepen our understanding of the stability and architecture principles of proteins.

REFERENCES

1. Charlier, D., and L. Droogmans. 2005. Microbial life at high temperature, the challenge, the strategies. *Cell Mol Life Sci* 24: 2974–84.
2. Brock, T. D., K. M. Brock, R. T. Belly, and R. L. Weiss. 1972. Sulfolobus: A new genus of sulfur-oxidizing bacteria living in low pH and high temperature. *Arch Microbiol* 84: 54–68.
3. Kashefi, K., and D. R. Lovley. 2003. Extending the upper temperature limit for life. *Science* 301: 934.
4. Vieille, C., and G. J. Zeikus. 2001. Hyperthermophilic enzymes: Sources, uses, and molecular mechanisms for thermostability. *Microbiol Mol Biol Rev* 65: 1–43.

5. Saiki, R. K., D. H. Gelfand, S. Stoffel, S. J. Schaarf, R. Higuchi, G. T. Horn, K. B. Mullis, and H. A. Erlich. 1998. Prime-directed enzymatic amplification of DNA with a thermostable DNA polymerase. *Science* 239: 487–91.
6. Barany, F. 1991. Genetic disease detection and DNA amplification using cloned thermostable ligase. *Proc Natl Acad Sci USA* 84: 6663–67.
7. Pace, C. N., E. J. Hebert, K. L. Shaw, D. Schell, V. Both, D. Krajcikova, J. Sevcik, K. S. Wilson, Z. Dauter, R. W. Hartley, and G. R. Grimsley. 1998. Conformational stability and thermodynamics of folding of ribonuclease Sa, Sa2 and Sa3. *J Mol Biol* 279: 271–86.
8. Jaenicke, R., and G. Böhm. 1998. The stability of proteins in extreme environments. *Curr Opin Struct Biol* 8: 738–48.
9. Kumar, S., C. J. Tsai, and R. Nussinov. 2000. Factors enhancing protein thermostability. *Protein Eng* 13: 179–91.
10. Kumar, S., and R. Nussinov. 2001. How do thermophilic proteins deal with heat? *Cell Mol Life Sci* 58: 1216–33.
11. Razvi, A., and J. M. Scholtz. 2006. Lessons in stability from thermophilic proteins. *Protein Sci* 15: 1569–78.
12. Pace, C. N. 1990. Measuring and increasing protein stability. *Trends Biotechnol* 8: 93–98.
13. Ruiz-Sanz, J., V. V. Filimonov, E. Christodoulou, C. E. Vorgias, and P. L. Mateo. 2004. Thermodynamic analysis of the unfolding and stability of the dimeric DNA-binding protein HU from the hyperthermophilic eubacterium *Thermotoga maritima* and its E34D mutant. *Eur J Biochem* 271: 1497–507.
14. Clark, A. T., B. S. McCrary, S. P. Edmondson, and J. W. Shriver. 2004. Thermodynamics of core hydrophobicity and packing in the hyperthermophile proteins Sac7d and Sso7d. *Biochemistry* 43: 2840–53.
15. Ge, M., X. Y. Xia, and X. M. Pan. 2008. Salt bridges in the hyperthermophilic protein Ssh10b are resilient to temperature increases. *J Biol Chem* 283: 31690–96.
16. McCrary, B. S., S. P. Edmondson, and J. W. Shriver. 1996. Hyperthermophile protein folding thermodynamics: Differential scanning calorimetry and chemical denaturation of Sac7d. *J Mol Biol* 264: 784–805.
17. Li, W. T., R. A. Grayling, K. Sandman, S. Edmondson, J. W. Shriver, and J. N. Reeve. 1998. Thermodynamic stability of archaeal histones. *Biochemistry* 37: 10563–72.
18. Grättinger, M., A. Dankesreiter, H. Schurig, and R. Jaenicke. 1998. Recombinant phosphoglycerate kinase from the hyperthermophilic bacterium *Thermotoga maritima*: Catalytic, spectral and thermodynamic properties. *J Mol Biol* 280: 525–33.
19. Shiraki, K., S. Nishikori, S. Fujiwara, H. Hashimoto, Y. Kai, M. Takagi, and T. Imanaka. 2001. Comparative analyses of the conformational stability of a hyperthermophilic protein and its mesophilic counterpart. *Eur J Biochem* 268: 4144–50.
20. Deutschman, W. A., and F. W. Dahlquist. 2001. Thermodynamic basis for the increased thermostability of CheY from the hyperthermophile *Thermotoga maritima*. *Biochemistry* 40: 13107–13.
21. Lee, C. F., M. D. Allen, M. Bycroft, and K. D. Wong. 2005. Electrostatic interactions contribute reduced heat capacity change of unfolding in a thermophilic ribosomal protein L30e. *J Mol Biol* 348: 419–31.
22. Razvi, A., and J. M. Scholtz. 2006. A thermodynamic comparison of HPr proteins from extremophilic organisms. *Biochemistry* 45: 4084–92.
23. Gianni, S., Y. Ivarsson, P. Jemth, M. Brunori, and C. Travaglini-Allocatelli. 2007. Identification and characterization of protein folding intermediates. *Biophys Chem* 128: 105–13.
24. Luke, K. A., C. L. Higgins, and P. Wittung-Stafshede. 2007. Thermodynamic stability and folding of proteins from hyperthermophilic organisms. *FEBS J* 274: 4023–33.

25. Mukaiyama, A., K. Takano, M. Haruki, M. Morikawa, and S. Kanaya. 2004. Kinetically robust monomeric protein from a hyperthermophile. *Biochemistry* 43: 13859–66.
26. Mukaiyama, A., M. Haruki, M. Ota, Y. Koga, K. Takano, and S. Kanaya. 2006. A hyperthermophilic protein acquires function at the cost of stability. *Biochemistry* 45: 12673–79.
27. Dong, H., A. Mukaiyama, T. Tadokoro, Y. Koga, K. Takano, and S. Kanaya. 2008. Hydrophobic effect on the stability and folding of a hyperthermophilic protein. *J Mol Biol* 378: 264–72.
28. Takano, K., R. Higashi, J. Okada, A. Mukaiyama, T. Tadokoro, Y. Koga, and S. Kanaya. 2009. Proline effect on the thermostability and slow unfolding of a hyperthermophilic protein. *J Biochem* 145: 79–85.
29. Mukaiyama, A., Y. Koga, K. Takano, and S. Kanaya. 2008. Osmolyte effect on the stability and folding of a hyperthermophilic protein. *Proteins* 71: 110–18.
30. Crouch, R. J., and M. L. Dirksen. 1982. Ribonuclease H. In *Nuclease*, vol. 14, ed. S. M. Linn and R. J. Robert, 211–41. Cold Spring Harbor, NY: Cold Spring Harbor Laboratory Press.
31. Ohtani, N., M. Haruki, M. Morikawa, and S. Kanaya. 1999. Molecular diversities of RNase H. *J Biosci Bioeng* 88: 12–19.
32. Muroya, A., D. Tsuchiya, M. Ishikawa, M. Haruki, M. Morikawa, S. Kanaya, and K. Morikawa. 2001. Catalytic center of an archaeal type2 ribonuclease H as revealed by X-ray crystallographic and mutational analyses. *Protein Sci* 10: 707–14.
33. Takano, K., S. Endo, A. Mukaiyama, H. Chon, H. Matsumura, Y. Koga, and S. Kanaya. 2006. Structure of amyloid beta fragments in aqueous environments. *FEBS J* 273: 150–58.
34. Takano, K., Y. Katagiri, A. Mukaiyama, H. Chon, H. Matsumura, Y. Koga, and S. Kanaya. 2007. Conformational contagion in a protein: Structural properties of a chameleon sequence. *Proteins* 68: 617–25.
35. Hollien, J., and S. Marqusee. 1999. A thermodynamic comparison of mesophilic and thermophilic ribonucleases H. *Biochemistry* 38: 3831–36.
36. Kaushik, J. K., K. Ogasahara, and K. Yutani. 2002. The unusually slow relaxation kinetics of the folding-unfolding of pyrrolidone carboxyl peptidase from a hyperthermophile, *Pyrococcus furiosus*. *J Mol Biol* 316: 991–1003.
37. Yamasaki, K., K. Ogasahara, K. Yutani, M. Oobatake, and S. Kanaya. 1995. Folding pathway of *Escherichia coli* ribonuclease H: A circular dichroism, fluorescence and NMR study. *Biochemistry* 34: 16552–62.
38. Raschke, T. M., J. Kho, and S. Marqusee. 1999. Confirmation of the hierarchical folding of RNase H: A protein engineering study. *Nat Struct Biol* 6: 825–31.
39. Hollien, J., and S. Marqusee. 2002. Comparison of the folding processes of *T. thermophilus* and *E. coli* ribonucleases H. *J Mol Biol* 316: 327–40.
40. Perl, D., C. Welker, T. Schindler, K. Schröder, M. A. Marahiel, R. Janicke, and F. X. Schmid. 1998. Conservation of rapid two-state folding in mesophilic, thermophilic and hyperthermophilic cold shock proteins. *Nat Struct Biol* 5: 229–35.
41. Schuler, B., W. Kremer, H. R. Kalbitzer, and R. Jaenicke. 2002. Role of entropy in protein thermostability: Folding kinetics of a hyperthermophilic cold shock protein at high temperatures using 19F NMR. *Biochemistry* 41: 11670–80.
42. Wallgren, M., J. Åden, O. Pylypenko, T. Mikaelsson, L. B.-Å. Johansson, A. Rak, and M. Wolf-Watz. 2008. Extreme temperature tolerance of a hyperthermophilic protein coupled to residual structure in the unfolded state. *J Mol Biol* 379: 845–58.
43. Pace, C. N. 1992. Contribution of the hydrophobic effect to globular protein stability. *J Mol Biol* 226: 29–35.
44. Takano, K., K. Ogasahara, H. Kaneda, Y. Yamagata, S. Fujii, E. Kanaya, M. Kikuchi, M. Oobatake, and K. Yutani. 1995. Contribution of hydrophobic residues to the stability of human lysozyme: Calorimetric studies and X-ray structural analyses of the five isoleucine to valine mutants. *J Mol Biol* 254: 62–76.

45. Takano, K., Y. Yamagata, S. Fujii, and K. Yutani. 1997. Contribution of hydrophobic effect to the stability of human lysozyme: Calorimetric studies and X-ray structural analysis of the nine valine to alanine mutants. *Biochemistry* 36: 688–98.

46. Takano, K., J. Funahashi, Y. Yamagata, S. Fujii, and K. Yutani. 1997. Contribution of water molecules in the interior of a protein to the conformational stability. *J Mol Biol* 274: 132–42.

47. Takano, K., Y. Yamagata, and K. Yutani. 1998. A general rule for the relationship between hydrophobic effect and conformational stability of a protein: Stability and structure of a series of hydrophobic mutants of human lysozyme. *J Mol Biol* 280: 749–61.

48. Matthews, B. W., H. Nicholson, and W. J. Beckel. 1987. Enhanced protein thermostability from site-directed mutations that decrease the entropy of unfolding. *Proc Natl Acad Sci USA* 84: 6663–67.

49. Watanabe, K., and Y. Suzuki. 1998. Protein thermostabilization by proline substitutions. *J Mol Catal B Enzyme* 4: 167–80.

50. Richardson, J. S., and D. C. Richardson. 1998. Amino acid preferences for specific locations at the ends of alpha helices. *Science* 240: 1648–52.

51. Yancey, P. H., and G. N. Somero. 1979. Counteraction of urea destabilization of protein structure by methylamine osmoregulatory compounds of elasmobranch fishes. *Biochem J* 18: 317–23.

52. Yancey, P. H., M. E. Clark, S. C. Hand, R. D. Bowlus, and G. N. Somero. 1982. Living with water stress: Evolution of osmolyte systems. *Science* 217: 1214–22.

53. Frye, K. J., and C. A. Royer. 1997. The kinetic basis for the stabilization of staphylococcal nuclease by xylose. *Protein Sci* 6: 789–93.

54. Russo, A. T., J. Rösgen, and D. W. Bolen. 2003. Osmolyte effects on kinetics of FKBP12 C22A folding coupled with prolyl isomerization. *J Mol Biol* 330: 851–66.

55. Okada, J., T. Okamoto, A. Mukaiyama, T. Tadokoro, D.-J. You, H. Chon, Y. Koga, K. Takano, and S. Kanaya. 2010. Evolution and thermodynamics of the slow unfolding of hyperstable monomeric proteins. *BMC Evol Biol* 10: 207.

56. Hickey, D. A., and G. A. C. Singer. 2004. Genomic and proteomic adaptations to growth at high temperature. *Genome Biol* 5: 117–121.

57. Musto, H., H. Naya, A. Zavala, H. Romero, F. Alvarez-Valin, and G. Bernardi. 2004. Correlations between genomic GC levels and optimal growth temperatures in prokaryotes. *FEBS Lett* 573: 73–77.

58. Elcock, A. H. 1998. The stability of salt bridges at high temperatures: Implications for hyperthermophilic proteins. *J Mol Biol* 284: 489–502.

59. Szilagyi, A., and P. Zavodszky. 2000. Structural differences between mesophilic, moderately thermophilic and extremely thermophilic protein subunits: Results of a comprehensive survey. *Structure* 8: 493–504.

60. Cambillau, C., and J. M. Claverie. 2000. Structural and genomic correlates of hyperthermostability. *J Biol Chem* 275: 32383–86.

61. Ree, D. C. 2001. Crystallographic analyses of hyperthermophile proteins. *Methods Enzymol* 334: 423–37.

62. Matsui, I., and K. Harata. 2007. Implication for buried polar contacts and ion pairs in hyperthermostable enzymes. *FEBS J* 274: 4012–22.

63. Subbian, E., Y. Yabuta, and U. P. Shinde. 2005. Folding pathway mediated by an intramolecular chaperone: Intrinsically unstructured propeptide modulates stochastic activation of subtilisin. *J Mol Biol* 347: 367–83.

64. Yabuta, Y., H. Takagi, M. Inoue, and U. P. Shinde. 2001. Folding pathway mediated by an intramolecular chaperone: Propeptide release modulates activation precision of prosubtilisin. *J Biol Chem* 276: 44427–34.

65. Yabuta, Y., E. Subbian, H. Takagi, U. P. Shinde, and M. Inoue. 2002. Folding pathway mediated by an intramolecular chaperone: Dissecting conformational changes coincident with autoprocessing and the role of Ca²⁺ in subtilisin maturation. *J Biochem* 131: 31–37.
66. Fu, X., M. Inoue, and U. P. Shinde. 2000. Folding pathway mediated by intramolecular chaperone. The inhibitory and chaperone functions of the subtilisin propeptide are not obligatorily linked. *J Biol Chem* 275: 16871–78.
67. Shinde, U. P., and M. Inoue. 2000. Intramolecular chaperones: Polypeptide extensions that modulate protein folding. *Semin Cell Dev Biol* 11: 35–44.
68. Inoue, M., X. Fu, and U. P. Shinde. 2001. Substrate-induced activation of a trapped IMC-mediated protein folding intermediate. *Nat Struct Biol* 8: 321–25.
69. Yabuta, Y., E. Subbian, C. Oiry, and U. P. Shinde. 2003. Folding pathway mediated by intramolecular chaperone. A functional peptide chaperone designed using sequence databases. *J Biol Chem* 278: 15246–51.
70. Falzon, L., S. Patel, Y. J. Chen, and M. Inoue. 2007. Automatic behavior of the propeptide in propeptide-mediated folding of prosubtilisin E. *J Mol Biol* 366: 494–503.
71. Eder, J., M. Rheinnecker, and A. R. Fersht. 1993. Folding of subtilisin BPN′: Role of the pro-sequence. *J Mol Biol* 233: 293–304.
72. Wang, L., S. Ruvinov, S. Strausberg, T. Gallagher, G. L. Gilliland, and P. N. Bryan. 1995. Prodomain mutations at the subtilisin interface: Correlation of binding energy and the rate of catalyzed folding. *Biochemistry* 34: 15415–20.
73. Wang, L., B. Ruan, S. Ruvinov, and P. N. Bryan. 1998. Engineering the independent folding of the subtilisin BPN′-pro-domain: Correlation of pro-domain stability with the rate of subtilisin folding. *Biochemistry* 37: 3165–71.
74. Kannan, Y., Y. Koga, Y. Inoue, M. Haruki, M. Takagi, T. Imanaka, M. Morikawa, and S. Kanaya. 2001. Active subtilisin-like protease from a hyperthermophilic archaeon in a form with a putative prosequence. *Appl Environ Microbiol* 67: 2445 552.
75. Pulido, M., K. Saito, S. Tanaka, Y. Koga, M. Morikawa, K. Takano, and S. Kanaya. 2006. Ca²⁺-dependent maturation of subtilisin from a hyperthermophilic archaeon, *Thermococcus kodakaraensis*: The propeptide is a potent inhibitor of the mature domain but is not required for its folding. *Appl Environ Microbiol* 72: 4154–62.
76. Tanaka, S., K. Saito, H. Chon, H. Matsumura, Y. Koga, K. Takano, and S. Kanaya. 2007. Crystal structure of unautoprocessed precursors of subtilisin from a hyperthermophilic archaeon. *J Biol Chem* 282: 8246–55.
77. Tanaka, S., H. Matsumura, Y. Koga, K. Takano, and S. Kanaya. 2007. Four new crystal structures of Tk-subtilisin in unautoprocessed, autoprocessed and mature forms: Insight into structural changes during maturation. *J Mol Biol* 372: 1055–69.
78. Takeuchi, Y., S. Tanaka, H. Matsumura, Y. Koga, K. Takano, and S. Kanaya. 2009. Requirement of a unique Ca²⁺-binding loop for folding of Tk-subtilisin from a hyperthermophilic archaeon. *Biochemistry* 48: 10637–43.
79. Tanaka, S., Y. Takeuchi, H. Matsumura, Y. Koga, K. Takano, and S. Kanaya. 2008. Crystal structure of Tk-subtilisin folded without propeptide: Requirement of propeptide for acceleration of folding. *FEBS Lett* 582: 3875–78.
80. Bryan, P. N., P. A. Alexander, S. L. Strausberg, F. Schwarz, W. Lan, G. L. Gilliland, and D. T. Gallagher. 1992. Energetics of folding subtilisin BPN′. *Biochemistry* 31: 4937–45.
81. Gallagher, D. T., P. N. Bryan, and G. L. Gilliland. 1993. Calcium-independent subtilisin by design. *Proteins* 16: 205–13.
82. Strausberg, S. L., P. A. Alexander, D. T. Gallagher, G. L. Gilliland, B. L. Barnett, and P. N. Bryan. 1995. Directed evolution of a subtilisin with calcium-independent stability. *Biotechnology* 13: 669–73.

83. Foophow, T., S. Tanaka, Y. Koga, K. Takano, and S. Kanaya. 2010. Subtilisin-like serine protease from hyperthermophilic archaeon *Thermococcus kodakaraensis* with N- and C-terminal propeptides. *Protein Eng Des Sel* 23: 347–55.

84. Foophow, T., S. Tanaka, A. Clement, Y. Koga, K. Takano, and S. Kanaya. 2010. Crystal structure of a subtilisin homologue, Tk-SP, from *Thermococcus kodakaraensis*: Requirement of a C-terminal β-jelly roll domain for hyperstability. *J Mol Biol* 400: 865–77.

85. Koga, Y., M. Morikawa, M. Haruki, H. Nakamura, T. Imanaka, and S. Kanaya. 1998. Thermostable glycerol kinase from a hyperthermophilic archaeon: Gene cloning and characterization of the recombinant enzyme. *Protein Eng* 11: 1219–27.

86. Yu, P., and D. W. Pettigrew. 2003. Linkage between fructose 1,6-bisphosphate binding and the dimer-tetramer equilibrium of *Escherichia coli* glycerol kinase: Critical behavior arising from change of ligand stoichiometry. *Biochemistry* 42: 4243–52.

87. Koga, Y., R. Katsumi, D.-J. You, H. Matsumura, K. Takano, and S. Kanaya. 2008. Crystal structure of highly thermostable glycerol kinase from a hyperthermophilic archaeon in a dimeric form. *FEBS J* 275: 2632–43.

2 Role of Packing, Hydration, and Fluctuations on Thermostability

Fabio Sterpone and Simone Melchionna

CONTENTS

2.1 INTRODUCTION

Proteins belonging to thermophilic organisms have the capacity to sustain high temperatures while maintaining a folded structure and preserving functionality. Understanding thermostability is one of the challenges regarding proteins that attract great attention from the scientific community. Thermostability is key to the comprehension of the crucial elements that keep proteins folded as well as determining the mechanisms that modulate stability. In the past, most research has focused on comparing mesophilic, thermophilic, and hyperthermophilic homologues (i.e., proteins belonging to the same family) with host organisms nearby in the philogenetic tree and sharing large portions of the amino acid sequence. A small difference of

5% of amino acid content may result in a shift in the melting temperature of tens of degrees Celsius.

After many years of studies in this field, several aspects have been identified as relevant to thermal resistance; others are still being debated. First, current research faces the often encountered dilemma of whether stability is induced by purely thermodynamic or by kinetic reasons. Next, the roles of specific interactions, such as van der Waals forces and electrostatics, have to be discerned, as well as quantifying their roles in specific hot spots of the protein matrix. Optimized packing of residues throughout the macromolecule is often advocated as the molecular mechanism strengthening the protein structure by enthalpic forces.

Many researchers have recently focused their attention on the role of fluctuations in thermostability. In analogy with the resistance of buildings against external agents, some researchers view thermal resistance as due to an entropic reservoir, as quantified by the enhancement of fluctuations with the degree of thermophilicity. Such an entropic reservoir would be able to soften the internal modes of macromolecules and protect them from thermal stress.

Finally, more recent research has focused on the role of hydration and the organization of the surrounding water in enhancing protein robustness. The formation of a collective network, or web of hydrogen bonds, would create a sufficient protecting envelope to sustain the protein scaffolding. Study of the morphological details of the protein–water interface points in this direction as well as the observation that melting of the protein–water hydrogen network acts as a precursor to protein unfolding.

These different aspects have a common denominator: they may all contribute to shifting the melting curve by a few degrees and thus their contribution is as small as a few kilocalories per mole. Moreover, they typically involve the global spatial arrangement of the protein by impacting the low-frequency region of the protein spectrum, slowing down the unfolding kinetics.

For all these reasons, experimental, theoretical, and computational methods need to be carefully chosen to provide the most effective scientific outcome in this fascinating sector of science.

This chapter focuses on computer simulation as the methodology of choice to investigate thermal resistance. Simulation has the capability to access the microscopic scale of proteins, revealing the motion of single atoms or providing a quantitative picture of the entire protein and its environment to the scale of hundreds of nanoseconds. The wealth of information that can be extracted is overwhelming. This information must be organized in a coherent way to provide a clear picture.

The picture of thermostability is still incomplete and requires deeper investigations looking at specific issues. The goal of this chapter is to provide an overview of the literature related to computational studies and compare the emerging ideas obtained from the joint use of simulations and experiments, with a particular eye on three putative molecular mechanisms for thermostability: packing, hydration, and fluctuations.

2.2   COMPUTER SIMULATION

Today computer modeling is an essential tool in the study of biomolecules. Simulations based on molecular dynamics (MD) and Monte Carlo techniques allow us to explore the phase space of biomolecules in equilibrium and out-of-equilibrium conditions, in different environments, and in different thermodynamic states. The ever-growing number of crystallographic structures deposited in protein data banks are analyzed aided by computer graphics and by models supporting research in structural biology: structural motifs are compared, docking virtual experiments are performed, *de novo* design and three-dimensional (3D) protein structure predictions are guided. Electrostatic calculations based on static structures represent an important aspect to gain insight into protein–protein interactions and to quantify macromolecular stability.

In the context of protein thermostability, several computational approaches are employed. Structural investigations are routinely carried out on the available structures of protein homologues from mesophilic, thermophilic, and hyperthermophilic organisms. The intramolecular contributions to thermal resistance are being investigated for homologue families and attention is being given to the specific role of salt bridges and intramolecular hydrogen bonds. MD simulations are typically performed to shed light on the relationship between protein stability and dynamics. The main results of several investigations will be discussed later in this chapter. In this section, we describe how modern computer simulations are used to investigate thermal resistance.

Molecular dynamics is a computer simulation method widely employed to study proteins and other biomolecules in solution. The guiding principle of MD is the direct determination of the classical motion of individual atoms, as encoded by Newton's equations of motion, once the elementary interatomic forces are known. Given the fact that accurate force fields are currently available to reproduce microscopic forces in polypeptides (e.g., CHARMM,[1] AMBER,[2] GROMOS,[3] to mention just a few), the simulation approach offers the possibility to study proteins from a microscopic, bottom-up perspective, and, in principle, without additional approximations or other sources of inaccuracy other than the force field itself.* Starting from the atomic detail, MD allows us to inspect conformational quantities as well as their fluctuations, the distribution of momentum within the protein, volumetric data, the distribution and persistence of salt bridges, hydrogen bonds, hydrophobic contacts, the magnitude of electrostatic interactions, and the dynamics of secondary structures. In addition, simulation provides direct access to the solvent to very high accuracy. In particular, by following the individual water molecules, one can analyze the network of hydrogen bonds, the single-molecule dynamics, and their collective behavior. All these aspects have been found to have important implications in the study of thermophilic proteins.

A limitation of MD is the current technological capabilities of available computers and the fact that the longest accessible timescale typically falls below the microsecond scale. In the literature, it is frequently found that proteins in water solution or embedded in lipid membranes are studied from a few up to tens or hundreds of nanoseconds, depending on the size of the system and the problem being investigated. In recent years, specialized parallel machines and high-performance software have

---

* In reality, some other approximations and modeling aspects intervene, such as the treatment of covalent bonds as either rigid or flexible entities, some numerical errors due to the finite timestep propagation of atomic trajectories, and the neglect of quantum effects, such as those involved in hydrogen bonding.

been deployed to stretch the simulation times.[4] In addition, new hardware based on the use of graphics cards employed as coprocessors is receiving increased attention from the community of biosimulations in order to substantially speed up calculations at affordable costs.

Actually, the timescale limitation precludes the direct observation of large-scale protein rearrangements, such as the unfolding or refolding processes. These phenomena can only be analyzed for small model polypeptides and by indirect methods. In this regard, it is worth mentioning that systematic work on developing simplified models for biomolecules has been recently undertaken by several groups (see Tozzini[5] and references therein). In the so-called coarse-grain approach, for example, the number of degrees of freedom of a system is reduced by representing the atoms belonging to a chemical group as a unique center of force (e.g., the side chain of an amino acid is described as a single grain). These types of models are used to explore processes occurring for times longer than the microsecond timescale, and examples include protein folding, aggregation and large-scale fluctuation, and membrane fusion and dynamics. Another strategy for bridging the time gap is to use an implicit representation of the solvent environment,[6] such as water or a membrane, or of the saline environment, such as via the Poisson–Boltzmann level of theory, but artifacts can be introduced by these types of approximation.

For analogous reasons, a large theoretical effort has been devoted to the development of methods to determine the free energy curves involving barriers much greater than the thermal energy, and whose spontaneous transitions occur beyond the nano- or microsecond timescale. In this scenario, the rare transitions between stable states can still be investigated by computational techniques that sample the conformational space of macromolecules in stepwise fashions, that is, by restraining the system to sample a given macrostate at one time. A plethora of methods[7] are available, beginning with the well-known umbrella sampling technique.[8] Sampling methods have become increasingly sophisticated, including metadynamics,[9] milestoning,[10] temperature accelerated MD,[11] replica exchange MD,[12] and reconstructing the reaction path between known stable states as in the string[89] and transition path sampling[14] methods.

There is no general approach for investigating protein stability that can cope with irreversible transitions, a condition where equilibrium thermodynamics is inapplicable, or when the number of microstates associated with the unfolded state is enormous and hard to sample in full. The sampling procedure becomes even more troublesome if the variable controlling stability is a state variable, such as temperature, and a mechanical reaction coordinate cannot be defined. This scenario also applies to the study of thermophilic proteins, with the further complication that when comparing mesophilic and thermophilic counterparts, the free energy differences are marginal and as small as a handful of van der Waals interactions. This renders the quest for the microscopic basis of thermostability akin to the needle in the haystack problem.

Simulation methods based on the full-fledged microscopic picture present further challenges. For instance, determination of the melting temperature by direct simulation is precluded, as it may take microseconds to observe complete destructuring of the protein scaffolding. Even observing the presence of an intermediate structure before complete unfolding occurs, as signaled by a plateau in the evolution of some structural quantity, is challenging. Instead, precise determination of the melting temperature would require many extended simulations of proteins prepared with different microscopic initial conditions, as in the folding@home project,[15] or the use of ad hoc procedures, such as the Replica Exchange MD method.[12] It is also worth noting that the melting temperature obtained in silico may be shifted with respect to the experimentally determined one, as pointed out recently.[4]

An effective approach is to observe the onset of protein melting by looking at steady trends in given observables. Several precursors of unfolding can be identified when temperature increases, such as the reduction of secondary structure elements, the gradual removal of hydrogen bonds and salt bridges, and the sudden enhancement of internal fluctuations. The identification of "good" indicators of stability is still challenging. When comparing mesophilic and thermophilic counterparts, one should not take the absolute amplitude of fluctuations as indicators of an incipient transition or as an indicator of protein stability. Rather, for a given species, the way fluctuation amplitudes vary with temperature should be taken as independent indicators of the ongoing unfolding process.

2.3 THERMODYNAMIC STABILITY

When dealing with thermophilic proteins, the basic question is to understand the physical principles, either thermodynamic or kinetic, that produce an enhanced thermal resistance at high temperatures and how these principles can be applied to mesophilic and thermophilic counterparts. The possibility of working with highly homologous proteins from mesophilic, thermophilic, and hyperthermophilic species offers a systematic way to understand the microscopic origin of stability starting at the molecular level. A global understanding of thermostability will enable us to assess the role of evolution in selecting the most effective and functional proteins in their host organisms, as well as design new, highly stable engineered biocatalysts.

Protein stability is, strictly speaking, measured by the difference in free energy between the folded and unfolded formations in equilibrium conditions. Kinetic stability depends on the rate of unfolding, which in turn depends on the barrier separating the folded and unfolded states. In many circumstances, the rate of unfolding correlates with thermodynamic stability, as a more stable protein is likely to present a greater barrier for unfolding. As sketched in Figure 2.1, one assumes that the free energy profile obeys a two-state model (i.e., with only two dominant states, folded and unfolded) and that proteins unfold reversibly in such a way as to apply equilibrium thermodynamics. In fact, if a protein unfolds irreversibly or follows complex unfolding pathways, a quantitative thermodynamic analysis is generally impossible.

When looking at protein stability as a function of a state variable, such as temperature, the free energy profile is globally modified in a way to alter the content of folded or unfolded structures in a statistical sense. From the balance of populations in the folded and unfolded states, one determines the free energy difference:

$$\Delta G = G(U) - G(F) = -K_B T \ln (<U>/<F>)$$

where <U> and <F> are the populations of the respective F and U states, and their variation with temperature.

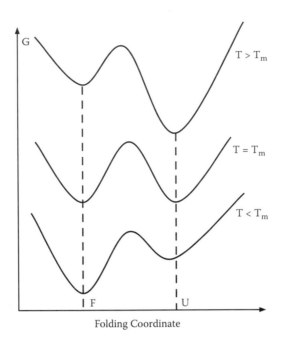

Folding Coordinate

FIGURE 2.1 Free energy profiles of a protein viewed according to the two-state model, illustrating the folded (F) and the unfolded (U) states separated by a simple free energy barrier. The three profiles are plotted according to a folding coordinate that quantifies the degree of order of the protein with respect to the native (ordered) state. The three curves represent profiles for low temperature ($T < T_m$), high temperature ($T > T_m$), and the melting (transition) temperature ($T = T_m$).

The typical modification of the curves with temperature produces the so-called stability curve, as sketched in Figure 2.2. The curve is the shape of a skewed inverted parabola, with a maximum at the temperature of maximum stability, T_s. The melting temperature, T_m, corresponds to an equal population of the F and U states, and for $T > T_m$, the protein is in the unfolded state. Even if there are mesophilic proteins that present high thermal stability, as in the case of bovine pancreatic trypsin inhibitor, which denatures at 373 K, the stability of thermophilic and hyperthermophilic species is generally higher, as T_m can reach the boiling temperature of water or higher.

The melting temperature of a thermophile can then be shifted to higher or lower values by modifying the protein characteristics in a number of ways. Thermostable proteins present a shift of T_m to higher values, and three possible modifications of the free energy profile can be envisaged. These are called the up-shift, right-shift, and broadening of the stability curves[16] (see Figure 2.2).

The way the stability curve is altered when going from the mesophilic to the thermophilic species reflects the modifications of the free energy profiles versus the folding coordinate. It has been observed that thermophilic proteins follow one of three strategies. Following a recent comparison of thermodynamic data on a dataset of 26 proteins,[17] it was shown that the majority of thermophiles cause an up-shift in the

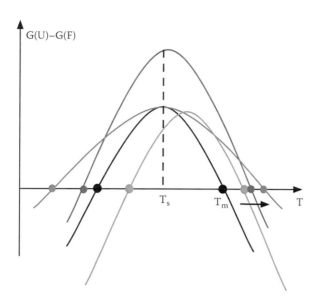

FIGURE 2.2 Stability curve of a given mesophilic protein (black curve) and possible modifications for the thermophilic homologue: up-shift (red curve), right-shift (orange curve), and broadening (magenta curve) of the stability curve. The two intercepts of each parabola with the axis correspond to equal populations of the folded and unfolded states and correspond to low (cold unfolding) and high (hot unfolding) temperatures. The intercepts are indicated with filled circles and the rightwise shift of the melting temperature, T_m, is indicated by the arrow. **(See color insert.)**

stability curve (77%) while others (31%) act to right-shift the stability curve by rigidly moving T_s to higher values. It was also shown that thermophiles often present a lower heat capacity of unfolding (70%) and therefore fall into the class of broadened free energy profiles. In general, thermophiles seem to choose any of these strategies or a combination thereof. It has been suggested that no single molecular mechanism can be ascribed to thermophilic proteins and most probably a mixture of stabilizing causes are the basis of thermal resistance.

Having said this, it is also plausible that the up-shift and right-shift of the stability curve are due to optimized intraprotein interactions that are capable of enhancing the structural robustness in a rather insensitive way from temperature so that a single molecular cause can be isolated in this respect. On the other hand, the often observed smaller heat capacity of unfolding in thermophiles suggests that more complex mechanisms are the basis of thermostability.

When analyzing the various proposals for interpreting thermal adaptation, one should keep in mind that the overall free energy difference between the denatured and native states of proteins is as small as 0.1 kcal/mol per residue, and the overall stability of the native state can be achieved by a surplus of few hydrogen bonds (2 to 4 kcal/mol each, even if the net gain depends on the difference with respect to the unfolded state where the groups are solvated and hydrogen bonded to water). As a result, several concomitant mechanisms may be capable of shifting the unfolding temperature by a few tens of degrees, and these can have an intraprotein origin or be related to solvation. These mechanisms may have a weak impact on the general structure or response of the protein to external agents, but have sufficient impact on shifting the denaturation temperature. In fact, it is estimated that increasing the denaturation temperature by 50 K involves a change in the free energy of unfolding by only a few kilocalories per mole.[18]

Calorimetric measurements are capable of probing the free energy landscape by measuring the heat capacity with respect to a reference state as a function of temperature. By working at constant pressure, the heat capacity C_p is defined as $C_p = dH/dT$, encoding the increase in enthalpy (H) with temperature (T). This equation can be rewritten in terms of the Gibbs free energy G as $C_p = -Td^2G/dT^2$. Therefore thermodynamics tells us that the heat capacity is proportional to the curvature of the stability curve versus temperature, with a positive (negative) change in C_p implying a more curved (flat) profile. In addition, the heat capacity has an equivalent definition, being proportional to the square fluctuations of the enthalpy, $C_p = <\delta H^2>/kT^2$, providing some insight into the protein fluctuations versus the folding coordinate.

If a protein is subjected to an external perturbation of some type (mutagenesis, pH, pressure, etc.), the heat capacity is consequently modified and from its sign one can infer important information. For mesophilic and thermophilic proteins, the unfolding ΔC_p is found to be positive[19] (with thermophiles exhibiting a smaller variation than mesophiles). By neglecting intraprotein contributions, this feature can be rationalized by considering, for instance, the contribution arising from apolar versus polar solvation.[20] A positive ΔC_p is a signature of dominant hydrophobic interactions, as in the case of globular proteins, whereas the solvation of polar groups in proteins has a negative sign.

An important question arises about which components contribute most to the measured heat capacity, in particular by isolating those arising from protein–protein contacts or from hydration. It is thought that hydration is dominant.[20] In different studies, it has been proposed that both hydration of polar groups and fluctuations of nonbonded interactions within the protein matrix play a dominant role upon unfolding. However, large effects are also thought to be produced cooperatively by the intramolecular network of weak dispersive interactions. The issue stands as an open question.

From a thermodynamic point of view, the heat capacity of unfolding measured by calorimetry has shown that thermophilic species often exhibit absolute smaller values, that is, the stability curve is flatter and shallower than in mesophiles. If the unfolded state has the same heat capacity for both the mesophilic and thermophilic counterparts, the smaller ΔC_p arises from the smaller curvature in the native state of the thermophile as compared to the mesophile.

By relying on a two-energy model, the heat capacity is interpreted on the basis of the magnitude of the enthalpic fluctuations rather than on enthalpy itself, suggesting that enthalpic fluctuations play a relevant role in thermostability.[20] A possibility is that the smaller ΔC_p in thermophiles arises from a wider and smoother free energy landscape as a function of the folding coordinate, as illustrated in Figure 2.3. However, a direct association of these fluctuations with molecular mechanisms or molecular fluctuations is still missing. The fundamental reason is that ΔC_p involves both entropic and enthalpic contributions and assessing which one prevails cannot be done on general grounds.

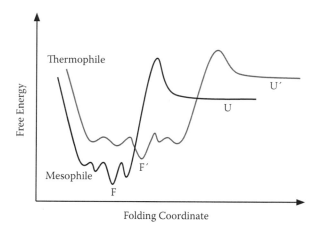

FIGURE 2.3 Pictorial representation of the free energy landscape of mesophilic and thermophilic homologues versus the folding coordinate. In this representation, the free energy basin of the thermophilic species is broader than its mesophilic counterpart. **(See color insert.)**

The supposed flatter landscape of thermophiles is a good match with the findings in the low-frequency, collective motions of the protein matrix, a part of the power spectrum that is difficult to access by experiments and simulations. The broader landscape relates to the fact that thermophiles sample more conformational states than the mesophilic analogues. In thermophiles, more numerous conformational states are somehow less tightly bound to the native state and the protein wanders more freely among several local basins.

An alternative view on the flatter free energy landscape in thermophiles was presented based on studies of ribonucleases H from *Thermus thermophilus* and *Escherichia coli*.[21] These proteins have a high homology but still show a marked difference in their change in heat capacity upon unfolding. It was proposed that in the unfolded state the difference in ΔC_p originates from the presence of a residual structure in the form of hydrophobic clusters in the thermophilic species. Hydrophobic cluster formation implies burial of an apolar surface, and consequently a decrease in heat capacity, as also verified by mutagenesis. The implication is that the measured specific heat does not really relate to complete unfolding, but rather to long-range interactions and cooperative processes in the denatured states that are responsible for the persistence of such clusters.

Unfolding studies are of particular relevance, as they shed light on the dilemma of whether stabilization has thermodynamic or kinetic origins. In early studies it was suggested that, at least in proteins like rubredoxin, thermostability is induced by kinetic trapping rather than thermodynamic stabilization.[22] Calorimetry has been applied to observe the irreversible denaturation of rubredoxin,[23] and variations of pH and ionic strength have shown that the stability of hyperthermophiles decreases as pH decreases. By looking at the unfolding process of thermophilic and mesophilic counterparts[22] at various pH conditions, it was noticed that (1) the modes associated with the unfolding process of thermophiles are softer than in mesophiles, and (2) electrostatics is not a unique factor affecting the stability of the homologues.

2.3.1 STRUCTURE AND FUNCTION

The structures of homologous proteins from thermophilic and hyperthermophilic organisms are generally similar to their mesophilic counterparts (i.e., with a root mean square difference in the main chain between rubredoxin from *Pyrococcus furiosus* and *Clostridium pasteurianum* of approximately 0.5 Å).[24] The structural similarity is a direct consequence of the high sequence identity, which is found to be in the 45% to 65% range.

Despite the structural similarity, thermophilic enzymes feature little activity under mesophilic conditions.[25,26] In order to explain such reduced catalytic power at comparable temperatures, and assuming that protein activity is mainly controlled by conformational flexibility, one infers that thermophilic enzymes are characterized by reduced dynamics at ambient conditions. The flexibility required for the enzymatic activity would then be recovered only at higher temperatures. Activity data further suggest that mesophilic and thermophilic enzymes behave similarly at their respective working temperatures. This concept, also known as the corresponding state model, was first introduced by Somero.[27]

Following this general idea, researchers have attempted to individuate the structural elements that differentiate (hyper)thermophiles from mesophiles and could be the origin of the supposed reduced mobility of (hyper)thermophiles at ambient conditions.

2.3.2 RIGIDITY AND PACKING

Thermostability is associated with the enhanced internal rigidity of the protein scaffolding. This picture emerged from amide proton and hydrogen–deuterium exchange rates, as well as resistance to proteolysis.[28]

At the basis of this concept is the role of enthalpic forces that are retained to enhance the robustness of the protein scaffolding. The forces responsible for such structural strengthening originate from van der Waals interactions of tightly packed nonpolar side chains. As an example, mutational analysis between histones from mesophilic and hyperthermophilic archaea revealed that the much greater stability of the hyperthermophilic counterparts is conferred predominantly by improved intermolecular hydrophobic interactions near the center of the histone dimer core and by additional favorable ion pairs on the dimer surface.[29] In addition, from the crystal structure of citrate synthase from the hyperthermophilic archaeon *P. furiosus*, an organism living at 373 K, and structural comparisons between the same enzyme from organisms growing optimally at 310 K (pig) and 328 K (*Thermoplasma acidophilum*), it has been shown that there is an increased compactness of the hyperthermophilic enzymes, a more intimate association of the subunits, an increase in the number of intersubunit ion pairs, and a reduction in thermolabile residues.[30] Recently a theoretical network analysis of the protein fold of rubredoxin suggested that the resistance to thermal stress of the hyperthermophilic protein is guaranteed by its mechanical rigidity.[31]

According to this picture, thermal robustness is associated with structural strength, modulated by the balance of intraprotein attractive and repulsive (steric) forces. The better packed interiors of thermophilic proteins leads to lower overall flexibility of the protein matrix and a corresponding reduction in the surface-to-volume ratio. Compactness may be achieved by shortening a number of loops, increasing the number of atoms buried from the solvent, optimizing the packing of side chains in the interior, and an absence of cavities.[32] Nevertheless, there is no general consensus, and a recent comparative structural study indicates that the internal packing does not distinguish substantially between mesophiles, thermophiles, and hyperthermophiles.[33]

At present, no general conclusion can be drawn: proteins working at extreme conditions seem to accumulate very different structural stabilizing elements, including a surplus of charge clusters, extended networks of hydrogen bonds, optimized internal packing, and enhanced hydrophobic contacts.[34,35]

2.4 FLUCTUATIONS

Following the corresponding state concept, higher rigidity and structural robustness have long been taken as synonyms of thermostability, with enthalpy-driven structural robustness being the leading cause of reduced internal flexibility. One would expect that some structural patterns or recurrences would show up at the secondary or tertiary structural level, but no significant secondary structure or pattern of salt bridges has been identified in association with thermostability. In addition, there is no evidence that better packing or extended hydrophobic contacts reduce internal fluctuations in complex macromolecular systems.

More recently, the paradigm that thermostability correlates with the suppression of internal fluctuations has been questioned by several experimental and simulation studies.

Rubredoxin from *P. furiosus*, the most thermostable globular protein currently known, was studied by hydrogen exchange experiments, demonstrating that conformational opening for solvent access occurs in the millisecond timescale ubiquitously over the protein matrix. The flexibility of the protein is sufficient for water to access the exchanging amide, thus the conclusion is that conformational fluctuations of the magnitude of the hydrogen bond rupture occur throughout the protein.[36] The main observation of this study regards the unexpected fluctuations in a hyperthermophilic protein, again implying that enhanced rigidity and thermal stability are not necessarily correlated. As a way out to the rigidity–flexibility conundrum, it is thought that different strategies may be adopted by different thermophilic proteins to improve stability.[35]

A study of thermophilic and mesophilic analogues of α-amylase via the joint use of infrared spectroscopy monitoring hydrogen/deuterium (H/D) exchange kinetics and incoherent neutron scattering measuring picosecond dynamics confirmed a similar free energy of stabilization of the two proteins at room temperature.[37] The thermophilic species was found to have greater flexibility compared with its mesophilic counterpart. Nuclear magnetic resonance (NMR) spectroscopy of the folding kinetics of a hyperthermophilic cold shock protein showed that the difference in conformational stability between hyperthermophilic (TmCsp) and mesophilic species (BsCspB) is solely due to the unfolding rate constant, underscoring the role of entropic stabilization of the thermophilic relative to its mesophilic counterpart.[38]

Several computational studies on mesophilic and thermophilic proteins have been performed by looking at the unfolding kinetics, in particular, by using high temperature to induce melting. Early MD studies on mesophilic proteins have shown that protein unfolding might be initiated at sites that are prone to large thermal fluctuations.[39] By analyzing the unfolding process of rubredoxin at rather high temperature (500 K) in mesophilic and hyperthermophilic counterparts, it was found that both proteins behave very similarly, with both proteins unfolding by first opening the loop region and then exposing the hydrophobic core.[40]

Simulation of two thermostable variants of para-nitrobenzyl esterase with 97% sequence identity to the wild type, and selected experimentally via in-laboratory evolution, showed that the thermostable variants remain closer to their crystal structures than the wild type while at the same time they display increased fluctuations about their time-averaged structures, suggesting that thermostability results in a suppression of large deviations from the native state and an increase in smaller-scale fluctuations around the native state. The thermophilic mutants show increased populations of the lowest-frequency modes, indicating that the thermally stable variants experience increased concerted motions relative to the wild type.[41]

More recent computer simulations of mesophilic and hyperthermophilic rubredoxins have shown that the latter has greater flexibility at temperatures of 300 K and 373 K, and that the overall flexibility of both enzymes at their optimal growth temperature is about the same. Interestingly, the conformational space sampled by both proteins was found to be larger at lower temperature.[42]

Recent simulations of the G domain of the elongation factor Tu (EF-Tu) protein from mesophilic, thermophilic, and hyperthermophilic organisms showed that protein flexibility is in general comparable over a wide range of temperatures. In some cases, flexibility increases with the degree of thermostability.[43,44] Similar results have recently been found for thermophilic and mesophilic flavoenzymes of the nitroreductase fold, simulated using a wide range of temperatures and for times reaching 50 ns.[45]

Given that each experimental or computational technique probes different temporal and spatial scales, ranging from the atomic level up to the movements of several groups of amino acids, the concept of flexibility has to be critically reconsidered. Generally speaking, several forms of flexibility can be identified in proteins, in particular by considering the frequency associated with the different fluctuations and relative scales. Therefore it is inappropriate to advocate a single measure of flexibility in a complex macromolecule. A protein can be rigid on a microscopic timescale and flexible on a longer timescale on the order of milliseconds or greater. In addition, flexibility/rigidity may be differently distributed over the protein matrix and may involve different modes.

The observation of similar or, in some cases, larger fluctuations in thermophiles indicates that it is not the enthalpic, but rather the entropic content that lay at the basis of the enhanced stability. In fact, rigidity does not offer any thermodynamic advantage for structural robustness except for simple systems, having nothing to do with the complexity of large biomolecules.

As calorimetric and infrared spectroscopic studies have suggested, the mechanism for increasing thermal stability is entropic stabilization; that is, the motion of some internal degree of freedom increases the internal entropy content. Given the fact that entropy contributes a negative term to the free energy, overall the system is stabilized. Computer simulations allow us to probe the fluctuations of the protein-occupied space by analyzing volumetric quantities and measuring compressibility, which is given by $\beta = \langle \delta V^2 \rangle / kT \langle V \rangle$, where the averages are taken in isothermal-isobaric conditions. The volume appearing in this equation involves effects of the protein's intrinsic compressibility and the surrounding hydration water. However, changes in the interstitial volumes dominate the contribution from the protein, which typically features a low density within the core, and the protein's intrinsic compressibility is the main contribution to the measured compressibility. Atom-based, volumetric analysis of mesophilic, thermophilic, and hypethermophilic EF-Tu G domains made with the Voronoi construction have shown that thermostability is related to larger intrinsic compressibility.[43,44] When compared with the mesophilic homologue, volume fluctuations in the hyperthermophilic and thermophilic G domains are comparable or even higher, and the atomic volume is slightly smaller. This finding indicates that the thermophile features a weakly better internal packing and, at the same time, a more fluid character—two concepts that may seem contradictory.

It is clear that identifying the molecular basis for the variation in compressibility with thermostability is of central importance. In this respect, Dadarlat and Post[46] introduced an interesting model, the so-called adhesive/cohesive model, that relates compressibility, protein stability, and the distribution of charged amino acid in the protein matrix. This model is discussed in detail in Section 2.5.

2.5   ELECTROSTATICS AND FUNCTION

Genomic analysis and direct investigation of protein structures have clearly shown that proteins from (hyper)thermophilic organisms are enriched in charged amino acids.[32,47,48] Hence, broadly speaking, electrostatic interactions have been considered a common trait of thermostability. Moreover, the large amount of charged amino acids correlates with the abundance of salt bridges in (hyper)thermophilic proteins. While in mesophilic proteins the number of salt bridges for 100 residues is about four to five on average, in hyperthermophiles this number rises to eight to nine.[49]

Since the early work of Perutz and Raidt[50] on ferrodoxin and hemoglobin A2, salt bridges and intramolecular hydrogen bonds have been considered crucial factors for stabilizing the protein fold and guaranteeing resistance to high temperature. In wet-lab experiments, site-direct mutagenesis has allowed researchers to explore how the creation of specific ion pair interactions in the protein matrix induces increased thermal stability. For example, in a seminal paper, Vetriani et al.[51] showed that in the case of a multisubunits protein, the hexameric glutamate dehydrogenases, the protein from the more thermophilic organism (*P. furiosus*), is characterized by an extended network of ion pairs at the subunit interface. This network is less extended in the less stable protein from *Thermococcus litoralis* because of the lack of a single salt bridge. Site-directed mutagenesis guided by homology modeling reintegrated the missing salt bridge and optimized the geometry of the surrounding hydrogen bond interactions. As a result, the kinetic thermal stability of the less stable protein had a fourfold increase.

2.5.1 Quantifying the Electrostatic Contribution

In order to quantify the contribution of electrostatic interactions to protein thermostability, systematic computational studies have been carried out on a number of protein families, each including members from mesophiles and (hyper)thermophiles.

Xiao and Honig[52] used an implicit solvent model and Poisson–Boltzmann modeling of the saline solvent to compute the electrostatic energy and the electrostatic contribution to the folding free energy of four protein families. Within the implicit representation of the solvent, their results indicate that the electrostatic contribution to the folding free energy is the common feature that differentiates thermophilic from mesophilic proteins. With the former being stabilized by the electrostatic contribution more than the latter, the electrostatic free energy gap between hyperthermophiles and mesophiles is evaluated in the range of −20 to −3 kcal/mol, depending on the family and the value of the dielectric constant used for the protein environment in the calculation. This contribution may or may not correlate with the number of salt bridges or hydrogen bonds detected in the protein fold. Indeed, depending on the protein structure, the ionizable groups are more or less exposed to the solvent, and hence are associated to a different desolvation free energy penalty. For the ferrodoxin and CheY families, the thermostable character of the proteins correlates to a lower desolvation free energy. In order to compensate for the desolvation penalty, charged groups totally or partially buried in the interior of a protein must be involved in a network of local electrostatic interactions (salt bridge and hydrogen bonds); it is the delicate balance between the network extension and the degree of solvent accessibility of these groups that makes the electrostatic contribution a stabilizing factor. It is worth noting that only when the electrostatic contribution to the free energy is computed does one appreciate the differences between hyperthermophiles, thermophiles, and mesophiles. The sole evaluation of electrostatic interactions is generally insensitive to the different thermostability of proteins. It is interesting to note that more favorable thermodynamic stability may result from a higher free energy of the unfolded state, as appears to be the case for the thermophilic *Bacillus caldolyticus* cold shock protein with respect to the mesophilic variant, according to recent theoretical calculations.[53]

2.5.2 Salt Bridges and Kinetic Stability

The networking of charged groups not only plays a role in a thermodynamics sense, but, depending on the location in the protein fold, may provide an important contribution to the kinetic barrier separating the folded and unfolded states. As previously mentioned, unfolding experiments supported by electrostatic calculations based on the Poisson–Boltzmann model indicated that salt bridges in rubredoxins may be the reason for the slower unfolding kinetics observed for the hyperthermophile compared with the mesophile.[22] According to the authors, salt bridges act as mild surface clamps, reducing and modifying the protein vibrational modes accessible by temperature. This funnel of thermal energy may be viewed as a "dissipative" process that protects the protein structure. However, it is still unclear how and where the excess thermal energy flows.

Protection may be reserved to a specific region of the protein, as discussed by Kumar et al.[54] Those authors investigated the structure and energetics of the salt bridge network of glutamate dehydrogenase using the Poisson–Boltzmann representation of a solvent medium. The clustering of charged groups around the active site appears as a peculiar feature of the thermophilic homologue compared with the mesophilic one. The authors concluded that, due to ion pair networking, the active site structure is preserved and kinetically resists the temperature increase, thus preserving the strategic region of the protein fold and its enzymatic function.

Using MD, Missimer et al.[55] investigated the effect of temperature on the occupancy of salt bridges in a ccβ-p helix in a monomer state and in a trimer assembly. Interestingly, the coil of three helices exhibits higher thermal stability correlated with long-life intra- and intermonomer salt bridges than for the monomer case. In spite of the entropy increase due to thermal excitation, the higher configurational disorder is accommodated stochastically by the network of salt bridges to preserve the global structure of the trimer. The larger extension of the salt bridge network, a consequence of intermonomer linking, globally confers robustness to the trimeric complex.

In order to understand the role of salt bridges in stability and kinetics, computational studies have specifically focused on the problem of ion pair interactions. MD simulations have been used to construct the potential of mean force (PMF) of ion group pairs[56] and charged amino acids.[57]. Thomas and Elcock[57] evaluated the probability of association of two charged amino acids (lysine and glutamate) and derived prototype molecules in a water solution and extracted the PMF from long-lasting simulations. The PMF as a function of the distance between charged terminals showed interesting features: (1) the existence of two minima, the first at a short distance mirroring the formation of a salt bridge, while the second corresponds to a solvent-separated contact; and (2) a more favorable association at all distances as temperature increases. These studies illustrate the modulation of the dielectric constant of the solvent and its lowering as temperature rises, rendering the salt bridge more stable from an energetic viewpoint.

Similarly, but using a biasing procedure to evaluate the PMF along a preferential direction, the work of Masunov and Lazaridis[56] focused on all possible ion pair interactions between amino acids and compared results from all MD simulations in explicit solvent with implicit solvent calculations. The strongest salt bridge interaction is formed between arginine and glutamate (−4.5 kcal/mol), and two minima are reproduced for many ion pair PMFs. It is difficult to compare the associated free energy barriers obtained in the two mentioned works, given the different methods employed and the different definitions in the reaction coordinate. In this respect, it would be interesting to evaluate the kinetic barrier in a protein environment.

MD simulations on thermophilic and mesophilic homologues[44,45,58] showed that, depending on their location, salt bridges are resistant to temperature increases. However, due to the limited simulation time (1 to 50 ns depending on the literature under consideration), it is hard to extract meaningful statistics on the kinetic process.

2.5.3 OPTIMIZED ELECTROSTATICS

As pointed out by Xiao and Honig,[52] the contribution of electrostatic energy to protein stability is not strictly correlated with the presence of localized salt bridges.

Indeed, besides the local environment (networking or favorable polar interactions), long-range effects should be considered as well.

By analyzing protein structures, it was found that thermodynamic stability is achieved mainly by minimizing unfavorable electrostatic interactions between like charges. In order to mimic and further optimize the evolutionary strategy, in recent years it has been proposed to "play" with electrostatics to design novel proteins with increased thermostability.[59–63]

Sanchez-Ruiz and Makhatadze[60] implemented a strategy for creating mutants with increased thermostability via surface charge–charge optimization. Redesign of charge–charge interactions on a protein surface appears to be a valuable strategy for increasing the electrostatic contribution to the folding/unfolding free energy because it hardly affects the protein folds, as mutations are localized on the surface and do not cause steric clashes, and even a small gain of a few kilocalories per mole may result in a major improvement of protein thermostability (folding/unfolding free energy differences are on the same order of magnitude).

The ingredients of the proposed protocol are the following: (1) an efficient algorithm to generate mutations and perform the optimization procedure and (2) a suitable model for computing the electrostatic contribution to folding/unfolding free energy for each generated mutant. To solve the first issue, a genetic algorithm was used[64] and, due to its simplicity and low computational cost, the Tanford–Kirkwood model was employed to estimate free energy differences. In this model, the protein is represented as a low-dielectric sphere immersed in a high-dielectric medium. Charges are then placed on the sphere according to the pairwise distance calculated from the 3D protein structure. The methodology was successfully applied and experimentally tested for enzyme stabilization: the designed mutants maintained the wild-type structure and their activity, and did not aggregate in solution as their thermostability was increased.[62] Electrostatic optimization may be achieved via charge addition, removal, or reversion. Cumulative addition of charges led to a saturation effect arising from the long-range nature of electrostatic interactions. It was further observed that at most eight substitutions are needed to gain free energy substantially.

While the strategy above focused on surface charges only, a recent investigation by Dadarlat and Post[46] suggested that protein stability depends on the in/out distribution of charges, where "in" indicates the buried charged groups and "out" indicates the distribution of charges at the protein surface. At first the authors found a correlation between the excess of charge at the surface and the compressibility of proteins. In protein where the majority of charges concentrate on the surface, increased compressibility was observed. When buried groups balance the solvent-exposed groups, and the excess of charge on the surface is reduced, compressibility decreases. This finding was rationalized via an adhesive/cohesive model, according to which cohesion of the protein fold is guaranteed by electrostatic interactions (salt bridges and hydrogen bonds) in the protein interior. Adhesion with the solvent is instead caused by the interaction between the charged group at the protein surface and interfacial water (see Figure 2.4).

The second finding, which is extremely pertinent in the context of protein stability, is that there is a correlation between the excess of charges on the surface and the heat capacity of unfolding. Again, a charge distribution strongly asymmetric

FIGURE 2.4 Pictorial view of the electrostatic interaction in the protein matrix and distribution of the charged amino acids in the interior of the protein and at the interface with water. The charged residues are represented as red (negatively charged as Asp and Glu) and blue (positively charged as Lys and Arg) spheres. When buried in the interior of the protein, the charged amino acids tend to form an extended network of ion pairs or hydrogen bonds in such a way as to compensate for the desolvation free energy penalty. At the protein surface, the charged groups form either strong hydrogen bonds with water or the ion pair. According to Dadarlat and Post,[46] the clustering at the interior of the protein gives rise to cohesive forces (arrows pointing to the interior of the core), while the distribution at the protein surface, via the coupling with interfacial water, favors adhesion to the solvent (arrows pointing toward the exterior). **(See color insert.)**

toward the surface of the protein is associated with a large heat capacity of unfolding, mirroring the minor stability of the protein. The adhesive/cohesive model has not been systematically tested against (hyper)thermophilic and mesophilic proteins, so the consequences on thermostability are unclear. In a recent study on thermostable homologues of the G domain of EF-Tu, the findings were not supportive of the predictions of the adhesive/cohesive model.[65] At a theoretical level, the cohesive contribution to the reduction in heat capacity was also considered by Zhou,[66] who showed that the presence of buried amino acids forming an extended network across the protein structure and internal hydrogen bond interactions may explain the reduced heat capacity of unfolding in thermophiles.[21]

In a recent work, Dominy, Minoux, and Brooks[67] analyzed the role of electrostatics on the stability of thermophilic proteins, but following a very different point of view. Their reasoning was based on calculation of the dielectric response of the protein matrix applying the Fröhlich–Kirkwood theory. By using MD to simulate members of two protein families, Csp and CheY, the authors found that the thermophilic homologues exhibit a higher dielectric constant. Because the dielectric constant is related to the fluctuations of the total dipole of the proteins, it is important to dissect which contribution prevails, the protein fluctuations or the charges of the proteins. The authors found that the protein matrix fluctuations are essentially the same between mesophilic and thermophilic species. Hence the higher dielectric constant associated with thermophiles is mainly caused by the charge-enriched protein surface. A higher internal dielectric constant clearly reduces the desolvation penalty and thus represents a favorable ingredient in protein stability. Thus the possibility to "locally" tune the dielectric constant of the protein matrix may represent a viable strategy for increasing thermostability. Mobile salt bridges at the protein surface provide a contribution in that direction.

2.5.4 ELECTROSTATICS: FLEXIBILITY AND ACTIVITY

To conclude, it is worth reviewing protein activity in relation to the character of thermophiles. This issue was discussed in the previous section when dealing with the problem of protein flexibility. We have reported that it is still unclear how the larger extent of charged amino acids of thermophilic proteins correlates with protein flexibility and stability. It is even more difficult to determine how the high density of charges correlates with the catalytic power of thermophilic enzymes.

By focusing on the reaction occurring at the protein active site, it has been shown that catalytic activity depends on the energy barrier separating the reactants from the products. Warshel[68] and Roca et al.[69] proved that enzymes preorganize their active site in order to reduce the reorganization free energy of the reaction. It is thus crucial to understand whether the large number of charged amino acids observed in thermophilic enzymes directly affect the free energy along the reaction path.

Salt bridges could act as stabilizers of the protein fold and then help preserve the optimal structure/electrostatics of the active site at high temperature. The increased activity with respect to the ambient temperature regime may be explained by considering the temperature dependence of the reaction rate. In this instance, the active site is organized in such a way that at ambient temperature the catalytic activity is strongly reduced due to the high activation energy, and activity is enhanced via a temperature increase. It is interesting to note that increasing temperature induces some structural changes in the protein matrix in general and in the active site in particular, an effect that could reduce the activation energy. Increasing temperature enhances the catalytic power of the enzyme even more. To date, and to the best of our knowledge, this problem has not been tackled in detail.

2.6 HYDRATION

In the previous sections, the role of an aqueous environment has been implicitly invoked. We recall here some of the known facts. The desolvation free energy penalty paid for burying charged amino acids, for instance, depends on the difference between the dielectric constant in the protein and in the solvent regions, and how such a difference changes with temperature.[70] Exposed ion pairs belonging to the protein surface may fluctuate between direct contact interaction and water-mediated

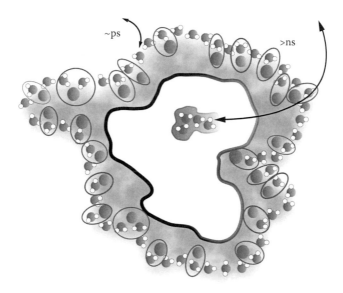

FIGURE 2.5 Organization of water around a protein. Water molecules form an extended network of hydrogen bonds surrounding the entire protein surface. The network is stabilized by local interactions with the hydrogen bond donor (blue spheres) and acceptor groups (red spheres) of the exposed amino acids. Depending on the composition of the protein surface, this layer is relatively stable with temperature. In the interior of the protein, polar or hydrophobic cavities may be filled with water molecules, and the water molecules are considered as structural elements of the protein matrix. While interfacial water molecules are quite mobile, with characteristic times slightly slower than bulk water molecules,[13,76] the internal water has reduced mobility and exchange with the bulk in nanoseconds or longer.[74,76] **(See color insert.)**

interaction, depending on the hydration of the charged amino acids. The flexibility/rigidity dilemma calls for a greater understanding of the role played by water in the stability and dynamics of proteins. It is largely accepted that protein and interfacial water dynamics are coupled at different timescales. Moreover, from both the kinetic and thermodynamic points of view, water is a principal actor in protein folding/unfolding.[71] For instance, pressure-induced unfolding is driven by water penetrating into the protein matrix that consequently destabilizes the hydrophobic core.[72,73]

In the following sections, the effects of water on protein thermostability are discussed. It is convenient to discuss separately the effects of so-called structural water versus the role of the interfacial hydration layer. The discussion is accompanied by a pictorial representation (see Figure 2.5).

2.6.1 INTERNAL WATER

For a long time, internal water has been considered a structural component of the protein matrix. Internal water molecules play a very specific structural action, for example, by creating molecular bridges and allowing extension of the electrostatic network. They also play a role in enzymatic function, for example, by altering the polarity of the binding site, and hence by enhancing the reactivity or creating a favorable highway for proton and ion mobility in the protein interior. Internal water molecules may act as an internal lubricant by favoring the relative displacement of secondary structure motifs. These molecules have been observed to exchange with bulk in the nanosecond timescale or more, a dynamic that correlates with the large-amplitude, breathing modes of proteins.[74–76]

Wetting of the protein interior depends on the local polarity of the host cavity,[77] and hydration occurs only cooperatively since a minimal number of water molecules is required for the formation of stable clusters. The penetration of water molecules in the protein interior is considered as the precursor to unfolding because of the hydrophobic interaction. However, if water locally satisfies its hydrogen-bonding requests with nearby amino acids, it can then access interior channels that represent functional paths to the protein.[78]

In the context of thermophilic proteins, Yin et al.[79] investigated wetting of the protein interior of the tetrabrachion stalk complex. They used MD simulations to estimate the free energy profile related to the wetting of cavities. The tetrabranchion stalk segment is a coiled coil assembly formed by four helices with a structure characterized by regular internal cavities of different size. Cavities are filled by water in the crystal structure determined at approximately 100 K. The authors addressed the question of whether the filling of the cavities is associated with stability and how the filling responds to increasing temperature. Two metastable states exist for the largest cavity, the first state corresponding to the empty cavity, and the second corresponding to a filled cavity with six to eight water molecules in a cluster. The two states are separated by a kinetic barrier that decreases with increasing temperature. The relative free energy between the two states changes as well, and the empty state is more stable at higher temperature. Therefore cavity drying is a precursor to protein denaturation and suggests that the filled state is key to protein stability. Interestingly, at the optimal growth temperature of 365 K, the two largest cavities approach the filling/emptying transition, a fact that probably affects the functional binding of proteases to the tetrabrachion stalk. Proteases recognize the empty cavities as a hydrophobic anchoring spot.

However, the role of internal water in the stability of the tetrabrachion stalk segment could be a specific case. It is worth noting that for a protein complex similar in structure (coiled coil formed by four α helices), the SNARE complex, the presence of water in internal cavities leads to reduced thermal stability of the *Saccharomyces cerevisiae* complex with respect to the neuronal one.[80]

A recent study on *in vitro* evolution of the lipase from mesophilic *Bacillus subtilis*[81] indicated that mutants with increased thermostability are characterized by specific interactions with water molecules. This happens by either extending the local hydrogen bond network via water bridging or by favoring the insertion of new amino acids via water displacement and replacement. At present, however, the role played by internal water in enhancing protein thermostability needs to be clarified. According to Pechkova et al.,[82] who performed a large study of 3D structures from the Protein Data Bank, it appears that thermophilic homologues bury less structural water than mesophilic homologues. The authors also remarked that the number of long-lived water molecules at the protein surface, as visible in crystal structures, are fewer and are less structured in thermophilic proteins.

2.6.2 THE WATER–PROTEIN INTERFACE

In a series of studies, MD simulations of the solvated G domain of EF-Tu from meso-philic, thermophilic, and hyperthermophilic organisms at different temperatures have explicitly focused on the behavior of water at the interface with proteins, with the aim of correlating hydration and thermostability.[43,44,65,83] When monitoring the protein surface exposed to water, for the first time a neat ordering between the three species was observed. In particular, the per atom contacts with water are maximized in hyperthermophiles and minimized in mesophiles.

Elementary thermodynamic considerations pointed out that the difference of the exposed surface is a key factor in explaining the higher stability of hyperthermo-philes, providing a contribution to the free energy stability larger or equivalent to that of internal packing. The extended water–protein contacts observed in (hyper) thermophiles correlate with a pronounced roughness of the protein surface. As a result, water accommodates well into the surface clefts and interstices and is highly coordinated with thermophiles.

The hydration layer of hyperthermophiles is also characterized by an increased number of solvent–protein hydrogen bonds. The hydrogen bond propensity compensates for the disruption of internal protein–protein hydrogen bonds and helps the protein preserve the fold as temperature increases.

Water appears to act as a local glue for (hyper)thermophiles, leading to enhanced thermostability. The protein surface of (hyper)thermophiles enriched in charged amino acids[65] anchors water at the surface, reducing the destabilizing process of deep water penetration. The hydration shell is more resistant to thermal increases in hyperthermophiles. When considering the global connectivity of water–water hydrogen bonds, it has been shown that hydration water around hyperthermophiles percolates around the protein surface, even at 330 K. At this temperature the hydration layer of the mesophile decomposes into disconnected small hydrogen-bonded clusters.

The modifications in the structural properties of hydration water, and in particular the percolation transition, come with or anticipate important conformational reorganizations of the protein matrix.[84] Indeed, sudden acceleration of protein dynamics is observed 10 K below the denaturation temperature.[85] The increase in the heat capacity[86] and H/D exchange[87] is initiated just prior to unfolding and generally mirrors a qualitative perturbation of the protein conformation. The percolation transition occurs when the adhesion of water to the protein surface, and the consequent local preferential orientation, is weak enough and, in order to satisfy the hydrogen bond potentiality, water strongly couples to the extrashell molecules rather than with intrashell molecules.

This was clearly shown by computing the heat capacity of water at the protein interface.[84] The same transition is seen when a rigid model of the protein surface is altered in order to increase its hydrophobic content; the transition is associated with typical dewetting and related structuring of water on a large hydrophobic surface area.[88] Therefore the stability of the hydration layer characterizing (hyper)thermophiles reflects how the specific composition of the protein surface enhances local coupling between the water hydrogen bond network and the protein. At the same time, composition limits the extension of single hydrophobic patches that may disrupt protein–water adhesion.[65]

2.7 CONCLUSION

Decoding the molecular origins of protein thermostability represents a breakthrough in understanding protein stability and paves the way to new uses of enzymes in biotechnology, chemistry, and pharmaceutics. This is a challenge, and modern computational methods provide an important toolbox for tackling the problem.

Simulations based on a detailed representation of proteins, on simplified coarse-grained models, or on exploiting advanced sampling techniques, carried out on massive parallel computers, provide high-throughput routes to investigate homologue proteins under diverse external conditions and under a wide range of temperatures. Sophisticated techniques for free energy calculation offer the possibility to investigate long-range structural rearrangements, such as ligand–protein binding, and hence to study how protein dynamics assists the process. Also, the joint use of computational techniques has shown how protein structures and dynamics enhance or reduce thermal resistance. Detailed investigations of protein–water coupling have shown the early stages of protein unfolding under thermal stress.

Determining the weak points of protein–water coupling serves as a guide for proposing mutations with the aim of modulating or tuning thermal stability. Along these lines, electrostatics plays a key role in modifying the free energy landscape and represents a relevant factor to enhance or reduce the stability of mutants.

In conclusion, recent advances in computational methods and newly available computational power have opened the door to new and insightful investigations in the field of protein thermostability.

REFERENCES

1. MacKerell, A. D., D. Bashford, M. Bellott, R. L. Dunbrack, J. D. Evanseck, M. J. Field, S. Fischer, J. Gao, H. Guo, S. Ha, D. Joseph-McCarthy, L. Kuchnir, K. Kuczera, F. T. K. Lau, C. Mattos, S. Michnick, T. Ngo, D. T. Nguyen, B. Prodhom, W. E. Reiher, B. Roux, M. Schlenkrich, J. C. Smith, R. Stote, J. Straub, M. Watanabe, J. Wiorkiewicz-Kuczera, D. Yin, and M. Karplus. 1998. All-atom empirical potential for molecular modeling and dynamics studies of proteins. *J Phys Chem B* 102(18): 3586–616.
2. Cornell, W. D., P. Cieplak, C. I. Bayly, I. R. Gould, K. M. Merz, D. M. Ferguson, D. C. Spellmeyer, T. Fox, J. W. Caldwell, and P. A. Kollman. 1995. A second generation force field for the simulation of proteins, nucleic acids, and organic molecules. *J Am Chem Soc* 117(19): 5179–97.
3. Oostenbrink, C., A. Villa, A. E. Mark, and W. F. van Gunsteren. 2004. A biomolecular force field based on the free enthalpy of hydration and solvation: The GROMOS force-field parameter sets 53a5 and 53a6. *J Comput Chem* 25(13): 1656–76.
4. Shaw, D. E., P. Maragakis, K. Lindorff-Larsen, S. Piana, R. O. Dror, M. P. Eastwood, J. A. Bank, J. M. Jumper, J. K. Salmon, Y. Shan, and W. Wriggers. 2010. Atomic-level characterization of the structural dynamics of proteins. *Science* 330(6002): 341–46.
5. Tozzini, V. 2005. Coarse-grained models for proteins. *Curr Opin Struct Biol* 15(2): 144–50.

6. Onufriev, A. 2008. Implicit solvent models in molecular dynamics simulations: A brief overview. In *Annual Reports in Computational Chemistry*, ed. R. A. Wheeler and D. C. Spellmeyer, 125–37. New York: Elsevier.

7. Frenkel, D., and B. Smit. 2002. *Understanding Molecular Simulation*. 2nd ed. London: Academic Press.

8. Torrie, G. M., and J. P. Valleau. 1977. Nonphysical sampling distributions in Monte Carlo free-energy estimation: Umbrella sampling. *J Comput Phys* 23(2): 187–99.

9. Laio, A., and M. Parrinello. 2002. Escaping free-energy minima. *Proc Natl Acad Sci USA* 99(20): 12562–66.

10. Faradjian, A. K., and R. Elber. 2004. Computing timescales from reaction coordinates by milestoning. *J Chem Phys* 120(23): 10880–89.

11. Maragliano, L., and E. Vanden-Eijnden. 2006. A temperature accelerated method for sampling free energy and determining reaction pathways in rare events simulations. *Chem Phys Lett* 426(1–3): 168–75.

12. Sugita, Y., and Y. Okamoto. 1999. Replica-exchange molecular dynamics method for protein folding. *Chem Phys Lett* 314(1–2): 141–51.

13. Marchi, M., F. Sterpone, and M. Ceccarelli. 2002. Water rotational relaxation and diffusion in hydrated lysozyme. *J Am Chem Soc* 124(23): 6787–91.

14. Bolhuis, P. G., D. Chandler, C. Dellago, and P. L. Geissler. 2002. Transition path sampling: Throwing ropes over rough mountain passes, in the dark. *Annu Rev Phys Chem* 53: 291–318.

15. Shirts, M., and V. S. Pande. 2000. Computing: Screen savers of the world unite! *Science* 290(5498): 1903–4.

16. Sterner, R. H., and W. Liebl. 2001. Thermophilic adaptation of proteins. *Crit Rev Biochem Mol* 36(1): 39–106.

17. Razvi, A., and J. M. Scholtz. 2006. Lessons in stability from thermophilic proteins. *Protein Sci* 15(7): 1569–78.

18. Rees, D. C., and M. W. Adams. 1995. Hyperthermophiles: Taking the heat and loving it. *Structure (Lond)* 3(3): 251–54.

19. Cooper, A. 2010. Protein heat capacity: An anomaly that maybe never was. *J Phys Chem Lett* 1(22): 3298–304.

20. Prabhu, N. V., and K. A. Sharp. 2005. Heat capacity in proteins. *Annu Rev Phys Chem* 56: 521–48.

21. Robic, S., M. Guzman-Casado, J. M. Sanchez-Ruiz, and S. Marqusee. 2003. Role of residual structure in the unfolded state of a thermophilic protein. *Proc Natl Acad Sci USA* 100(20): 11345–49.

22. Cavagnero, S., D. A. Debe, Z. H. Zhou, M. W. Adams, and S. I. Chan. 1998. Kinetic role of electrostatic interactions in the unfolding of hyperthermophilic and mesophilic rubredoxins. *Biochemistry* 37(10): 3369–76.

23. Klump, H. H., M. W. W. Adams, and F. T. Robb. 1994. Life in the pressure cooker: The thermal unfolding of proteins from hyperthermophiles. *Pure Appl Chem* 66(3): 485–89.

24. Day, M. W., B. T. Hsu, L. Joshua-Tor, J.-B. Park, Z. H. Zhou, M. W. W. Adams, and D. C. Rees. 1992. X-ray crystal structures of the oxidized and reduced forms of the rubredoxin from the marine hyperthermophilic archaebacterium *Pyrococcus furiosus*. *Protein Sci* 1(11): 1494–507.

25. Wrba, A., A. Schweiger, V. Schultes, R. Jaenicke, and P. Zavodszky. 1990. Extremely thermostable D-glyceraldehyde-3-phosphate dehydrogenase from the eubacterium *Thermotoga maritima*. *Biochemistry* 29(33): 7584–92.

26. Voelkl, P., P. Markiewicz, K. O. Stetter, and J. H. Miller. 1994. The sequence of a subtilisin-type protease (aerolysin) from the hyperthermophilic archaeum *Pyrobaculum aerophilum* reveals sites important to thermostability. *Protein Sci* 3(8): 1329–40.

27. Somero, G. N. 1978. Temperature adaptation of enzymes. *Ann Rev Ecol Syst* 9: 1–29.
28. Daniel, R. M., D. A. Cowan, H. W. Morgan, and M. P. Curran. 1982. A correlation between protein thermostability and resistance to proteolysis. *Biochem J* 207(3): 641–44.
29. Li, W. T., J. W. Shriver, and J. N. Reeve. 2000. Mutational analysis of differences in thermostability between histones from mesophilic and hyperthermophilic archaea. *J Bacteriol* 182(3): 812–17.
30. Russell, R. J. M., J. M. C. Ferguson, D. W. Hough, M. J. Danson, and G. L. Taylor. 1997. The crystal structure of citrate synthase from the hyperthermophilic archaeon *Pyrococcus furiosus* at 1.9 Å resolution. *Biochemistry* 36(33): 9983–94.
31. Rader, A. J. 2009. Thermostability in rubredoxin and its relationship to mechanical rigidity. *Phys Biol* 7: 16002.
32. Vieille, C., and G. J. Zeikus. 2001. Hyperthermophilic enzymes: Sources, uses, and molecular mechanisms for thermostability. *Microbiol Mol Biol Rev* 65(1): 1–43.
33. Karshikoff, A., and R. Ladenstein. 1998. Proteins from thermophilic and mesophilic organisms essentially do not differ in packing. *Protein Eng* 11(10): 867–72.
34. Jaenicke, R., and G. Böhm. 1998. The stability of proteins in extreme environments. *Curr Opin Struct Biol* 8(6): 738–48.
35. Jaenicke, R. 2000. Do ultrastable proteins from hyperthermophiles have high or low conformational rigidity? *Proc Natl Acad Sci USA* 97(7): 2962–64.
36. Hernandez, G., F. E. Jenney, M. W. W. Adams, and D. M. LeMaster. 2000. Millisecond timescale conformational flexibility in a hyperthermophile protein at ambient temperature. *Proc Natl Acad Sci USA* 97(7): 3166–70.
37. Fitter, J., and J. Heberle. 2000. Structural equilibrium fluctuations in mesophilic and thermophilic alpha-amylase. *Biophys J* 79(3): 1629–36.
38. Schuler, B., W. Kremer, H. R. Kalbitzer, and R. Jaenicke. 2002. Role of entropy in protein thermostability: Folding kinetics of a hyperthermophilic cold shock protein at high temperatures using 19F NMR. *Biochemistry* 41(39): 11670–80.
39. Daggett, V., and M. Levitt. 1992. A model of the molten globule state from molecular dynamics simulations. *Proc Natl Acad Sci USA* 89(11): 5142–46.
40. Lazaridis, T., I. Lee, and M. Karplus. 1997. Dynamics and unfolding pathways of a hyperthermophilic and a mesophilic rubredoxin. *Protein Sci* 6(12): 2589–605.
41. Wintrode, P. L., D. Zhang, N. Vaidehi, F. H. Arnold, and W. A. Goddard III. 2003. Protein dynamics in a family of laboratory evolved thermophilic enzymes. *J Mol Biol* 327(3): 745–57.
42. Grottesi, A., M.-A. Ceruso, A. Colosimo, and A. Di Nola. 2002. Molecular dynamics study of a hyperthermophilic and a mesophilic rubredoxin. *Proteins* 46(3): 287–94.
43. Melchionna, S., R. Sinibaldi, and G. Briganti. 2006. Explanation of the stability of thermophilic proteins based on unique micromorphology. *Biophys J* 90(11): 4204–12.
44. Sterpone, F., C. Bertonati, G. Briganti, and S. Melchionna. 2009. Key role of proximal water in regulating thermostable proteins. *J Phys Chem B* 113(1): 131–37.
45. Merkley, E. D., W. W. Parson, and V. Daggett. 2010. Temperature dependence of the flexibility of thermophilic and mesophilic flavoenzymes of the nitroreductase fold. *Protein Eng Des Sel* 23(5): 327–36.
46. Dadarlat, V. M., and C. B. Post. 2003. Adhesive-cohesive model for protein compressibility: An alternative perspective on stability. *Proc Natl Acad Sci USA* 100(25): 14778–83.
47. Berezovsky, I. N., W. W. Chen, P. J. Choi, and E. I. Shakhnovich. 2005. Entropic stabilization of proteins and its proteomic consequences. *PLoS Comput Biol* 1(4): e47.
48. Tekaia, F., and E. Yeramian. 2006. Evolution of proteomes: Fundamental signatures and global trends in amino acid compositions. *BMC Genomics* 7: 307.

49. Karshikoff, A., and R. Ladenstein. 2001. Ion pairs and the thermotolerance of proteins from hyperthermophiles: A "traffic rule" for hot roads. *Trends Biochem Sci* 26(9): 550–56.
50. Perutz, M. F., and H. Raidt. 1975. Stereochemical basis of heat stability in bacterial ferredoxins and in haemoglobin a2. *Nature* 255(5505): 256–59.
51. Vetriani, C., D. L. Maeder, N. Tolliday, K. S. Yip, T. J. Stillman, K. L. Britton, D. W. Rice, H. H. Klump, and F. T. Robb. 1998. Protein thermostability above 100 degrees C: A key role for ionic interactions. *Proc Natl Acad Sci USA* 95(21): 12300–305.
52. Xiao, L., and B. Honig. 1999. Electrostatic contributions to the stability of hyperthermophilic proteins. *J Mol Biol* 289(5): 1435–44.
53. Zhou, H.-X., and F. Dong. 2003. Electrostatic contributions to the stability of a thermophilic cold shock protein. *Biophys J* 84(4): 2216–22.
54. Kumar, S., B. Ma, C. J. Tsai, and R. Nussinov. 2000. Electrostatic strengths of salt bridges in thermophilic and mesophilic glutamate dehydrogenase monomers. *Proteins* 38(4): 368–83.
55. Missimer, J. H., M. O. Steinmetz, R. Baron, F. K. Winkler, R. A. Kammerer, X. Daura, and W. F. van Gunsteren. 2007. Configurational entropy elucidates the role of salt-bridge networks in protein thermostability. *Protein Sci* 16(7): 1349–59.
56. Masunov, A., and T. Lazaridis. 2003. Potentials of mean force between ionizable amino acid side chains in water. *J Am Chem Soc* 125(7): 1722–30.
57. Thomas, A. S., and A. H. Elcock. 2004. Molecular simulations suggest protein salt bridges are uniquely suited to life at high temperatures. *J Am Chem Soc* 126(7): 2208–14.
58. Bae, E., and G. N. Phillips, Jr. 2005. Identifying and engineering ion pairs in adenylate kinases. Insights from molecular dynamics simulations of thermophilic and mesophilic homologues. *J Biol Chem* 280(35): 30943–48.
59. Mozo-Villarías, A., J. Cedano, and E. Querol. 2003. A simple electrostatic criterion for predicting the thermal stability of proteins. *Protein Eng* 16(4): 279–86.
60. Sanchez-Ruiz, J. M., and G. I. Makhatadze. 2001. To charge or not to charge? *Trends Biotechnol* 19(4): 132 35.
61. Schweiker, K. L., and G. I. Makhatadze. 2009. Protein stabilization by the rational design of surface charge-charge interactions. *Methods Mol Biol* 490: 261–83.
62. Gribenko, A. V., M. M. Patel, J. Liu, S. A. McCallum, C. Wang, and G. I. Makhatadze. 2009. Rational stabilization of enzymes by computational redesign of surface charge-charge interactions. *Proc Natl Acad Sci USA* 106(8): 2601–6.
63. Basu, S., and S. Sen. 2009. Turning a mesophilic protein into a thermophilic one: A computational approach based on 3D structural features. *J Chem Inf Model* 49(7): 1741–50.
64. Ibarra, B., and J. M. Sanchez-Ruiz. 2002. Genetic algorithm to design stabilizing surface-charge distributions in proteins. *J Phys Chem B* 106(26): 6609.
65. Sterpone, F., C. Bertonati, and S. Melchionna. 2010. Water around thermophilic proteins: The role of charged and apolar atoms. *J Phys Condens Matter* 22(28): 284113.
66. Zhou, H.-X. 2002. Toward the physical basis of thermophilic proteins: Linking of enriched polar interactions and reduced heat capacity of unfolding. *Biophys J* 83(6): 3126–33.
67. Dominy, B. N., H. Minoux, and C. L. Brooks, 3rd. 2004. An electrostatic basis for the stability of thermophilic proteins. *Proteins* 57(1): 128–41.
68. Warshel, A. 2003. Computer simulations of enzyme catalysis: Methods, progress, and insights. *Annu Rev Biophys Biomol Struct* 32: 425–43.
69. Roca, M., H. Liu, B. Messer, and A. Warshel. 2007. On the relationship between thermal stability and catalytic power of enzymes. *Biochemistry* 46(51): 15076–88.
70. Elcock, A. H. 1998. The stability of salt bridges at high temperatures: Implications for hyperthermophilic proteins. *J Mol Biol* 284(2): 489–502.

71. Levy, Y., and J. N. Onuchic. 2006. Water mediation in protein folding and molecular recognition. *Annu Rev Biophys Biomol Struct* 35: 389–415.

72. Hummer, G., S. Garde, A. E. García, M. E. Paulaitis, and L. R. Pratt. 1998. The pressure dependence of hydrophobic interactions is consistent with the observed pressure denaturation of proteins. *Proc Natl Acad Sci USA* 95(4): 1552–55.

73. Day, R., and A. E. García. 2008. Water penetration in the low and high pressure native states of ubiquitin. *Proteins* 70(4): 1175–84.

74. Sterpone, F., M. Ceccarelli, and M. Marchi. 2001. Dynamics of hydration in hen egg white lysozyme. *J Mol Biol* 311(2): 409–19.

75. García, A. E., and G. Hummer. 2000. Water penetration and escape in proteins. *Proteins* 38(3): 261–72.

76. Halle, B. 2004. Protein hydration dynamics in solution: A critical survey. *Philos Trans R Soc Lond B Biol Sci* 359(1448): 1207–23; discussion 1223–24, 1323–28.

77. Rasaiah, J. C., S. Garde, and G. Hummer. 2008. Water in nonpolar confinement: From nanotubes to proteins and beyond. *Annu Rev Phys Chem* 59: 713–40.

78. Scorciapino, M. A., A. Robertazzi, M. Casu, P. Ruggerone, and M. Ceccarelli. 2010. Heme proteins: The role of solvent in the dynamics of gates and portals. *J Am Chem Soc* 132(14): 5156–63.

79. Yin, H., G. Hummer, and J. C. Rasaiah. 2007. Metastable water clusters in the nonpolar cavities of the thermostable protein tetrabrachion. *J Am Chem Soc* 129(23): 7369–77.

80. Strop, P., S. E. Kaiser, M. Vrljic, and A. T. Brunger. 2008. The structure of the yeast plasma membrane snare complex reveals destabilizing water-filled cavities. *J Biol Chem* 283(2): 1113–19.

81. Ahmad, S., M. Z. Kamal, R. Sankaranarayanan, and N. M. Rao. 2008. Thermostable *Bacillus subtilis* lipases: In vitro evolution and structural insight. *J Mol Biol* 381(2): 324–40.

82. Pechkova, E., V. Sivozhelezov, and C. Nicolini. 2007. Protein thermal stability: The role of protein structure and aqueous environment. *Arch Biochem Biophys* 466(1): 40–48.

83. Melchionna, S., G. Briganti, P. Londei, and P. Cammarano. 2004. Water-induced effects on the thermal response of a protein. *Phys Rev Lett* 92(15): 158101.

84. Oleinikova, A., I. Brovchenko, and G. Singh. 2010. The temperature dependence of the heat capacity of hydration water near biosurfaces from molecular simulations. *EPL* 90(3): 36001.

85. Russo, D., J. Pérez, J.-M. Zanotti, M. Desmadril, and D. Durand. 2002. Dynamic transition associated with the thermal denaturation of a small beta protein. *Biophys J* 83(5): 2792–800.

86. Privalov, P. L., N. N. Khechinashvili, and B. P. Atanasov. 1971. Thermodynamic analysis of thermal transitions in globular proteins. I. Calorimetric study of chymotrypsinogen, ribonuclease, and myoglobin. *Biopolymers* 10(10): 1865–90.

87. Nakanishi, M., M. Tsuboi, and A. Ikegami. 1973. Fluctuation of the lysozyme structure. II. Effects of temperature and binding of inhibitors. *J Mol Biol* 75(4): 673–82.

88. Pizzitutti, F., M. Marchi, F. Sterpone, and P. J. Rossky. 2007. How protein surfaces induce anomalous dynamics of hydration water. *J Phys Chem B* 111(26): 7584–90.

89. Weinan, E., W. Ren, and E. Vanden-Eijnden. 2002. String method for the study of rare events. *Phys Rev B* 66(5): 052301.

3 Analyzing Protein Rigidity for Understanding and Improving Thermal Adaptation

Doris L. Klein, Sebastian Radestock, and Holger Gohlke

CONTENTS

3.1 INTRODUCTION

Investigating the thermal stability of proteins is of central importance for science and industry.[1] During the past few decades there has been considerable effort focused on understanding the structural determinants of protein stability, which is of interest for protein engineering and design as well as structure prediction of proteins.[1] In particular, the role of noncovalent interactions for the stabilization of a protein's native tertiary structure has become the focus of protein science.[2,3] Naturally occurring proteins from thermophilic organisms, referred to here as "thermophilic proteins," retain their native structures up to temperatures of 80°C while their homologue counterparts from mesophilic organisms, referred to here as "mesophilic proteins," denature.[4–8] Thermophilic proteins are valuable for investigating the molecular forces that determine thermostability. Such knowledge can be applied to improve the thermostability of proteins from mesophilic organisms, which is an important task in engineering proteins for biotechnological and chemical applications.[9] Comparative studies using homologues from mesophilic and thermophilic organisms have revealed that thermophilic adaptation often comes with better packing of hydrophobic interactions and an increased density of salt bridges or charge-assisted hydrogen bonds.[10–12] In many cases, a complex interplay of different sequences or structural features has been found to be responsible for increased thermostability.[13–15] As a unifying concept, it was suggested that these changes contribute to the improvement of the underlying network of noncovalent interactions within the structure, presumably leading to an overall increase in the mechanical stability/rigidity of the structure.[16] At the same time, an appropriate distribution of flexible regions must be maintained in the thermostable protein because biological function and molecular motions are intimately linked.[4,17]

For investigating the relations among protein sequence or structure, structural rigidity or flexibility, and thermostability, it is valuable to recall that rigidity and flexibility are concepts firmly grounded in mathematics,[18,19] solid-state physics,[20] and structural engineering.[21] Here, structural rigidity and flexibility denote static properties. No relative motion is allowed in a rigid (structurally stable) region; that is, a rigid structure does not bend or flex. In contrast, motion is possible within a flexible region; that is, this region can be deformed.[22–24] Promising computational approaches that allow determination of rigid and flexible regions within a protein structure have only recently been applied to investigate and improve the thermal stability of proteins.[25–28] These approaches will be described in more detail below.

Frequently the term flexibility is also used to describe mobility (movement, motion), although, in a strict sense, flexibility describes the *possibility* of motion. The distinction between mobility and flexibility becomes clear in the case of a rigid body that moves (e.g., a moving helix). Approaches for investigating structural determinants of thermostability by analyzing protein mobility are based on molecular dynamics (MD) simulation[29] and normal mode analysis (NMA).[30] By performing MD simulations at different temperatures or by running nonequilibrium simulations, motional properties of the protein during thermal unfolding can be investigated.[31–34] By comparing equilibrium and nonequilibrium MD simulations of cutinase, Creveld et al.[35] were able to predict unfolding regions in the protein where mutations influence thermostability but do not interfere with enzyme activity. Without doing expensive MD simulations, moving protein parts can be predicted from a single protein conformation using NMA.[30] Using a related approach, Burioni et al.[36] analyzed the effects of global topology on the thermal stability of folded proteins.

3.2 LINKING PROTEIN RIGIDITY AND THERMAL ADAPTATION

3.2.1 RIGIDITY THEORY

Analyzing the rigidity and flexibility of objects was already of scientific interest in the 18th and 19th centuries, when Lagrange[37] and Maxwell[21] began to study the mechanical stability of buildings or bridges. Maxwell's approach is based on the counting of constraints in a framework and relating the number of these constraints, C, to the number of degrees of freedom of the system. In the context of a bridge, the constraints are represented by steel rods that connect joints. The number of degrees of freedom can be determined by considering that each joint in a bridge has 3 degrees of freedom, resulting in $3n$ degrees of freedom for n joints, from which 6 degrees of freedom for the overall translation and rotation of the object are subtracted. In total, this yields

$$dof = 3n - 6 - C \qquad (3.1)$$

where dof is the number of internal independent degrees of freedom, also referred to as "floppy modes." If $dof = 0$, the object is rigid; otherwise it is flexible. While conceptually appealing, it should be noted that Maxwell's constraint counting is only valid if no redundant constraints exist in the framework.

More than a century later, a theorem by Laman[18] had a major impact on modern rigidity theory in that it allows us to precisely determine the degrees of freedom in a two-dimensional framework, even in the presence of redundant constraints, by applying constraint counting to all subgraphs within the framework. In this way the framework can be decomposed into rigid regions with flexible links in between. According to the molecular framework conjecture,[38,39] the constraint counting can be extended to a certain subtype of three-dimensional frameworks with a molecule-like character, the so-called bond-bending networks or molecular frameworks. Together with the development of efficient combinatorial algorithms to determine rigid and flexible regions, even in large frameworks,[40,41] this provided the opportunity to apply rigidity theory to a wide range of chemical and biological systems, including inorganic glasses,[42] zeolithes,[43] proteins,[24,44–47] and RNAs.[48,49]

3.2.2 MODELING AND ANALYZING PROTEINS AS CONSTRAINT NETWORKS

In order to perform a rigidity analysis on a protein structure, the structure is modeled as a molecular framework, also called a constraint network. In this network, vertices (joints) represent atoms, and edges (struts) represent bond constraints as well as angular constraints. Modeling covalent bonds is straightforward in that respect. However, given that the mechanical rigidity of biomacromolecules is largely determined by noncovalent interactions, the outcome of a rigidity analysis is governed by the way hydrogen bonds, salt bridges, and hydrophobic interactions are modeled in the network.[45] Here, hydrogen bonds and salt bridges are included as distance and angular constraints between the hydrogen bond donating and accepting groups (Figure 3.1).[24] The hydrogen bonds are ranked according to an energy function that

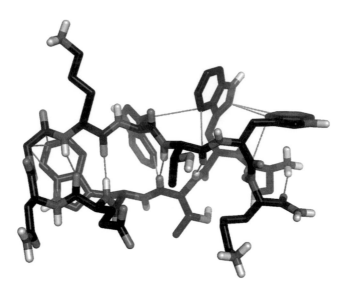

FIGURE 3.1 Constraint network of the tryptophan zipper 2 structure (PDB code 1le1). Covalent interactions are shown in black, hydrogen bonds are shown in magenta, and hydrophobic interactions are shown in green. Nitrogen atoms are colored in blue and oxygen atoms in red. Only polar hydrogen atoms are shown for clarity. **(See color insert.)**

takes into account the hybridization state of donor and acceptor atoms as well as their mutual orientation;[24,50] only hydrogen bonds with an energy (E_{hb}) below a given cutoff (E_{cut}) are retained in the network. Hydrophobic interactions are included if the distance between a pair of carbon or sulfur atoms is smaller than the sum of the atoms' van der Waals radii plus a threshold (Figure 3.1).

Based on previous studies that showed a negligible effect of structural water molecules on the rigidity of proteins[25] and a protein–protein complex,[45] water molecules are usually not considered when building the network. The same applies to molecules from the crystallization buffer, substrates, and cofactors. However, it is worth mentioning that such molecules could be included when investigating their effect on protein stability. In contrast, metal ions that are part of the protein structure are considered in general because of their large impact on protein stability.

The constraint network representation is analyzed with the FIRST[51] (Floppy Inclusion and Rigid Substructure Topography) software, which implements the pebble game algorithm[40,52] for determining local network rigidity. Remarkably, a FIRST analysis of a molecule of several thousand atoms takes just a few seconds, such that FIRST is also applicable to large biomacromolecules.[49,53–55]

3.2.3 THERMAL UNFOLDING SIMULATION BASED ON CONSTRAINT NETWORK ANALYSIS

In order to determine the melting temperature (T_m) of a protein as a descriptor of the protein's thermostability, thermal unfolding experiments are performed.[56] The increasing temperature weakens noncovalent interactions, destabilizing the protein structure. At the melting temperature, a transition from the folded to the unfolded state of the protein is observed. In an analogous way, thermal unfolding is simulated on a constraint network representation of a protein. Here, an increase in temperature is mimicked by gradually removing hydrogen bonds from the network in the order of increasing strength. This follows the idea that stronger hydrogen bonds will break at higher temperatures than weaker ones.[50] The number of hydrophobic contacts is kept constant during the thermal unfolding because the strength of hydrophobic interactions remains constant or even increases with increasing temperature.[57] Finally, a rigidity analysis is performed on the new constraint network.

In principle, the constraint network analysis (CNA)[27] can be performed on a single, static three-dimensional structure of a biomacromolecule. However, different conformations of a protein structure can lead to different results in the rigidity analysis, as observed by us[45] and others.[58] To overcome this problem, we pursued an ensemble-based approach in which conformations extracted from a trajectory generated by an MD simulation are individually subjected to CNA. The results from the thermal unfolding simulations are then averaged over the whole ensemble. As a further advantage over analyzing a single structure, this approach allows us to determine the statistical significance of the results of the CNA.

3.2.4 ANALYZING THERMAL UNFOLDING SIMULATIONS

At the beginning of a thermal unfolding simulation, the network representation of the protein contains many constraints. Rigidity analysis on this network shows that the protein is dominated by a large rigid cluster, also referred to as a giant cluster (Figure 3.2a). As more and more constraints are removed from the network, the giant cluster decays and large parts of the protein become flexible (Figure 3.2). This phase transition is not continuous. Rather, at the rigidity percolation threshold,[42] the giant cluster suddenly breaks and stops dominating the system. Interestingly, for materials as different as proteins and glasses, such a transition can be observed.[44] However, the percolation behavior of protein networks is usually more complex, and several transitions can be observed for these biomacromolecules. This is related to the fact that protein structures are modular, as they are assembled from secondary structure elements, subdomains, and domains. These modules often break away from the giant cluster as a whole. The first transition observed during the unfolding simulation describes the breakdown of the completely rigid network into a number of rigid clusters. The last transition describes the loss of the remaining piece of rigidity and the onset of a network that is completely flexible (i.e., rigid clusters cease to exist at this point). Thus the last transition is biologically most relevant because it corresponds to the folded–unfolded transition in experimental protein unfolding.[27] The temperature at which this transition occurs is referred to as the phase transition temperature (T_p). It can be related to the experimental melting temperature (T_m) (see below).

For identifying the transitions in thermal unfolding simulations, parameters originating from percolation theory[59] are applied.[27] The rigidity order parameter (P_∞) monitors the decay of the giant cluster by denoting the fraction of atoms of the constraint network that belong to the giant cluster. In the unfolded state, P_∞ approaches zero. As shown in Figure 3.3a for thermolysin-like protease (TLP), P_∞ provides an

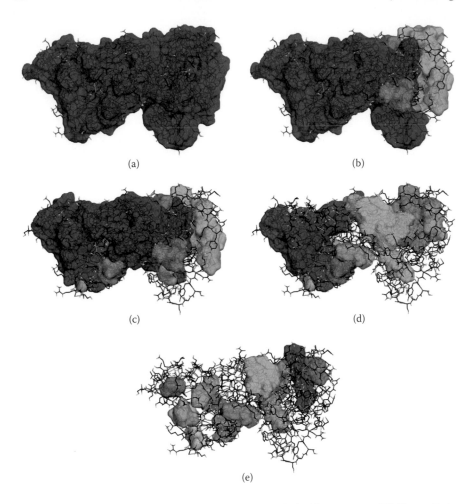

FIGURE 3.2 Rigid cluster decomposition of thermolysin-like protease (TLP), applied to constraint network representations of TLP at (a) 306, (b) 320, (c) 334, (d) 348, and (e) 362 K. Rigid clusters are shown as colored blobs, with the giant cluster colored in blue. Flexible regions are displayed as sticks. **(See color insert.)**

intuitive and detailed description of the residual rigidity within the protein structure during the unfolding process: with increasing temperature, the complex percolation behavior of the system is revealed by multiple, steep decreases of P_∞. In fact, different protein folds lead to different percolation behaviors because of the unique number and pattern of spatially arranged structural modules in each fold.

For identifying T_p, another parameter, the cluster configuration entropy (H)[60] has proved more valuable in our hands.[27] H analyzes macroscopic properties of the network that are associated with the rigid cluster size distribution[61] and has been adapted from Shannon's information theory. It measures the degree of disorder in a given state of the constraint network and is defined as

$$H = -\sum_{s} w_s \ln w_s \qquad (3.2)$$

where w_s is the probability that an atom belongs to a cluster of size s^2. As long as the rigid cluster dominates the system, H is low (Figure 3.3b). At the point where the giant cluster stops dominating the network, H rises abruptly, indicating the transition from the folded to the unfolded state of the protein. The temperature at which this jump occurs is considered the T_p.[27]

When considering microscopic properties of the constraint network before and after the transition, local regions in the protein structure at which the unfolding starts can be identified.[27] Residues that are part of the giant cluster before the transition, but are in a flexible region afterwards, constitute these unfolding nuclei or weak spots. Obviously, such regions are prominent candidates for introducing mutations that lead to higher thermostability. By pursuing an ensemble-based approach, a per residue probability for being part of an unfolding nucleus is obtained by averaging the results of thermal unfolding simulations over the ensemble.[62]

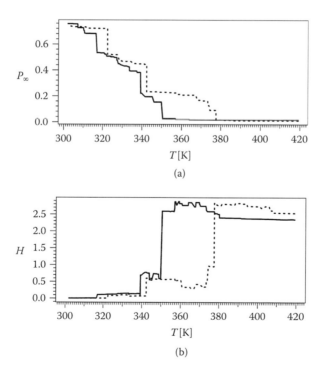

FIGURE 3.3 (a) Rigidity order parameter P_∞ and (b) cluster configuration entropy H versus temperature for mesophilic TLP (solid line) and thermophilic thermolysin (dashed line). (Adapted from Radestock, S., and H. Gohlke, 2011, Protein Rigidity and Thermophilic Adaptation, *Proteins* 79: 1089–1108.)

3.3 APPLYING CONSTRAINT NETWORK ANALYSIS FOR UNDERSTANDING AND IMPROVING THERMAL ADAPTATION

3.3.1 RETROSPECTIVE IDENTIFICATION OF UNFOLDING NUCLEI

Initially CNA was applied to identify structural features from which a destabilization of a protein structure originates upon thermal unfolding.[27] Structurally weak regions in proteins have been investigated before by experiment and computational studies. These regions have been called critical regions,[63] first unfolding regions,[35] and unfolding nucleation sites.[64]

When applied to mesophilic TLP from *Bacillus cereus* and thermophilic thermolysin, both of which belong to the family of M4 peptidases,[65] similar unfolding nuclei were identified by CNA. For both proteins, the unfolding nuclei are located in the β sheet of the N-terminal domain, a binding region of a calcium ion, a hydrophobic region around Phe63, and at the N-terminus of the α helix in the N-terminal domain (Figure 3.4). Notably, the predicted unfolding regions are in good agreement

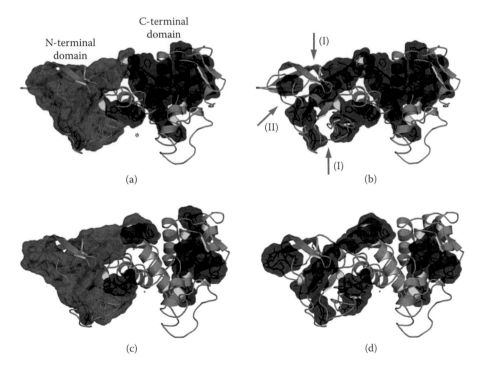

FIGURE 3.4 Snapshots from the thermal unfolding simulation of mesophilic TLP (a, b) and thermophilic thermolysin (c, d) just before (a, c) and after the phase transitions (b, d) at 350 and 373 K, respectively. The rigid cluster decomposition of the network is shown. The giant cluster is shown in blue. Other clusters are shaded in black. Arrows in (b) indicate potential unfolding nuclei. Roman numbers refer to the numbering of the unfolding nuclei. (*) The asterisk marks the active site. (Adapted from Radestock, S., and H. Gohlke, 2011, Protein Rigidity and Thermophilic Adaptation, *Proteins* 79: 1089–1108.) **(See color insert.)**

with sites where stabilizing mutations have been introduced by experiment into TLP, thereby increasing thermostability.[66–72]

When applied to 3-isopropylmalate dehydrogenase (IPMDH) from *Escherichia coli* and *Thermus thermophilus*—dimeric enzymes involved in the leucine biosynthesis pathway—the decay of the giant cluster causes residues at the domain interface, in domain 1, and at the subunit interface (domain 2) to become flexible. Again, sites within the predicted regions have also been successfully targeted by experiments with the aim of increasing the thermostability of mesophilic and chimeric IPMDH.[73–84] The experimental examination of thermophilic IPMDH has further revealed that most of the sites that are important for thermostability are at the subunit interface,[73,81,85] the domain interface,[76,86,88] and in domain 1.[89,90] These regions are also detected as unfolding regions by CNA for both the mesophilic and the thermophilic IPMDH.

Overall, good (albeit not perfect) agreement between predicted unfolding nuclei and thermostabilizing experimental mutations is encouraging, in that it demonstrates that CNA can be helpful to guide data-driven protein engineering to regions where mutations most likely should have a notable effect on thermostability. Equally important, the approach also identifies regions that must not be manipulated. Given the efficiency of the method, the approach can thus be used as a prefilter in a hierarchical computational scheme, followed by more sophisticated but also more expensive methods to evaluate the effect of an actual mutation on stability.

3.3.2 Proof of Concept: Improving the Thermal Stability of Phytase

In a prospective study, CNA was applied to identify unfolding nuclei in phytases. Subsequently this knowledge was used to guide mutagenesis experiments in an endeavor to increase the thermal stability of these proteins (S. Radestock, A. Shivange, U. Schwaneberg, S. Haefner, and H. Gohlke, unpublished results). Phytases are phytic acid–decomposing enzymes that are used as additives in green fodder for poultry and pigs. Here, phytases improve the availability of phosphorus, amino acids, and minerals, and help reduce the phosphorus excretion of monogastric animals.[91] The industrial process of generating the additive requires temperatures between 70°C and 90°C, which leads to high demands on the thermal stability of phytases. At the same time, phytases must remain active at the body temperature of the animals (30°C–40°C).

The wild-type phytase of *Yersinia mollaretii* was used for the identification of unfolding nuclei, applying an ensemble-based approach.[62] Three main unfolding nuclei were identified, which mainly comprise the β sheet and an α helix in the α/β domain. At two of the identified unfolding nuclei, saturation mutagenesis was performed, and the thermal stability of mutant enzymes was determined in terms of the temperature at which 50% of the enzyme molecules still show activity (T_{50}). When exchanging Val4 to Ala, T_{50} increased by 0.7°C compared with the wild-type enzyme. Albeit small, this increase is still remarkable because single mutations have rarely led to an increase of T_{50} by more than 1°C in the case of phytases.[92] On the other hand, exchanging Ala344 to Ile and Arg decreased T_{50} by 2.8°C and 2.3°C, respectively. This result is undesirable from the point of view of improving the thermal stability of *Y. mollaretii* phytase. Nevertheless, both results demonstrate as a proof of concept that mutations located in unfolding nuclei do have a pronounced effect on the protein's thermal stability. Clearly, more work is needed in this area, particularly with respect to predicting which mutations should or should not be introduced at an unfolding nucleus.

3.3.3 Toward Understanding Thermal Adaptation: Considering Enzyme Activity in Addition to Thermal Stability

In the retrospective study on identifying unfolding nuclei, we found strong agreement between structural events that take place at the unfolding transition in the mesophilic enzyme and its thermophilic homologue, for both TLP and IPMDH. In particular, the location and size of the giant cluster were similar, as were the identified unfolding regions (Figure 3.4). This finding is summarized in the P_∞ curves for TLP (Figure 3.3a), which characterize the general percolation behavior of the constraint networks during unfolding. With respect to the pattern of transitions, the P_∞ curve of the thermophilic protein is almost identical to its mesophilic counterpart except it is shifted to higher temperatures. From a global point of view, these observations provide initial support to the hypothesis of corresponding states,[4,93] according to which mesophilic and thermophilic enzymes are in corresponding states of similar rigidity and flexibility at their respective optimal temperatures.

In order to test this hypothesis more widely, we analyzed the local distribution of flexible and rigid regions in 19 pairs of homologous proteins from meso- and thermophilic organisms and related these distributions to activity characteristics of the enzymes.[94] For this, we compared microscopic stability features of the homologues with the help of stability maps, introduced for the first time in this study. The average correlation coefficient calculated over 19 pairs of stability maps of thermophilic and mesophilic homologues amounts to 0.53 ± 0.19 (SD), with a maximal correlation coefficient of 0.90, demonstrating that microscopic stability features of homologues agree well. Phrased differently, this comparison shows that adaptive mutations in enzymes from thermophilic organisms maintain the balance between overall rigidity, which is important for thermostability, and local flexibility, which is important for activity, at the appropriate working temperature.

Next, the pairs of TLP and IPMDH structures were analyzed with respect to the conservation of the distribution of functionally important flexible regions in the active site regions. In Figure 3.5, stability features of the mesophilic (Figure 3.5a,b) and thermophilic (Figure 3.5c,d) TLP active sites are shown at both the working temperature of the mesophilic (Figure 3.5a,c) and the thermophilic (Figure 3.5b,d) protein. Remarkably, at the proteins' respective optimal working temperatures (Figure 3.5a,d), an almost perfect agreement between the distribution of rigid and flexible regions of the active sites is revealed. In both cases, two of the three catalytic residues are placed on a flexible helix, while the third one is part of a rigid cluster. Furthermore, the close surroundings of the catalytic residues display highly similar stability features in both proteins. Such a remarkable correspondence of stability properties was also observed for the active sites of the IPMDH homologues at their respective optimal working temperatures.[94]

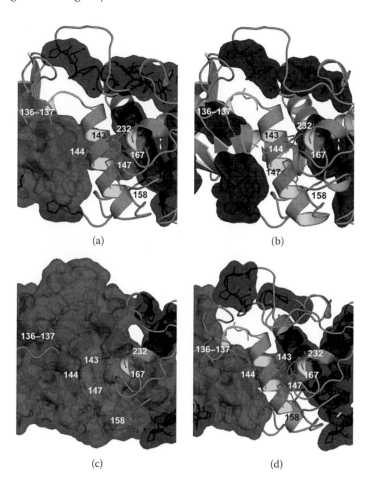

(a)   (b)

(c)   (d)

FIGURE 3.5 Active site of mesophilic TLP (a, b) and thermophilic thermolysin (c, d) at the working temperature of the mesophilic enzyme (342 K) (a, c) and the working temperature of the thermophilic enzyme (364 K) (b, d). The rigid cluster decomposition of the network is shown. The giant cluster is shown in blue. Other clusters are shaded in black. Catalytic residues are shown in red, while other functionally important residues are shown in green. (Adapted from Radestock, S., and H. Gohlke, 2011, Protein Rigidity and Thermophilic Adaptation, *Proteins* 79: 1089–1108.) **(See color insert.)**

Thus, changes in the flexibility of active site regions induced either by a temperature change or by a mutation were qualitatively related to experimentally observed losses of the enzyme function. First, at temperatures 20 to 30 K below T_p, the overall flexibility of the active site region of thermolysin was lost because of a now-rigid hinge region that is important for catalysis and substrate binding. Additional contacts, established between residues of the hinge region and residues of the neighboring giant cluster, were found to be responsible. This finding explains data from experiments that show almost no activity of thermophilic thermolysin at a temperature 15 K below the optimal temperature of the enzyme.[95] Second, mutational studies were undertaken to further analyze the role of the flexibility of the TLP hinge region.[70,96]

These studies revealed that mutating hinge residue Leu145 to Ser improved the enzyme's activity but did not influence thermostability.[96] CNA provides a rationalization for this observation in that it shows that the Leu145Ser mutation was indeed performed in a region that is not important for macroscopic stability. Overall, these analyses revealed that activity can be modulated independently from stability by introducing a mutation at a site that does not correspond to an unfolding region.

From an application point of view, these results demonstrate that exploiting the principle of corresponding states via CNA not only allows for successful optimization of thermostability, but also for guiding experiments in a qualitative manner in order to improve (or, at least, not deteriorate) enzyme activity in the course of protein engineering. An obvious next step is to derive a quantitative relationship between (local) changes of rigidity/flexibility and changes of enzyme activity, which is an area of active research in our laboratory.

3.4 OTHER APPROACHES BASED ON RIGIDITY THEORY

Using an approach related to CNA introduced by us,[27] Rader[28] investigated the thermal stability of a pair of homologous rubredoxin proteins from the mesophilic organism *Clostridium pasteurianum* and the hyperthermophilic organism *Pyrococcus furiosus* by applying rigidity theory. Rubredoxin is an electron-transferring protein in archaea and some anaerobic bacteria. As in CNA, a thermal unfolding was simulated by diluting the hydrogen bond network of the protein. The unfolding transition was determined either from the maximum in the function that relates the second derivative of the floppy mode density with respect to the mean coordination,[44] by the fraction of atoms in the largest rigid cluster, or from the steepest increase of the cluster configuration entropy, H. Here, the calculation of H follows the one in CNA (Equation 3.2) except that w_s now denotes the probability that an atom belongs to a cluster of size s, not s^2. Using the cluster size s leads to identifying an early percolation step during the thermal unfolding when the network-spanning giant cluster starts to decay. At this point, the protein loses its ability to carry stress. In contrast, using the cluster size s^2, as in CNA, leads to identifying a later percolation step when the giant cluster ceases to exist. At this point, the protein loses its residual structural stability; hence, in our opinion, the s^2-based cluster configuration entropy is better suited to identify unfolding nuclei whose stabilization improves overall protein stability. Along these lines, using the s^2-based cluster configuration entropy often identifies core regions as unfolding nuclei, while the s-based cluster configuration entropy results in the identification of loop regions or regions that are not part of the protein core.

For the rubredoxin example, Rader[28] found that the hyperthermophilic protein is structurally more rigid at a given temperature than the mesophilic one, in agreement with our CNA studies on other pairs of homologues. In addition, Rader investigated local rigidity properties of rubredoxin and compared them with protection factors of amide hydrogens ln$K$ derived from hydrogen–deuterium (H/D) exchange experiments. Such protection factors provide information about the structural rigidity of a residue because the most rigid residues are also the most protected and slowly exchanging ones. Similar information is provided by the propensity of a residue to be part of the largest rigid cluster, P_{lrc}. The differences between the respective results of the mesophilic and hyperthermophilic rubredoxin, $\Delta \ln K$ and ΔP_{lrc}, were then used to investigate which residues are more rigid in the hyperthermophilic protein than the mesophilic one. In both cases, residues in the β sheet and the metal-binding region were identified that are more rigid in the hyperthermophilic protein, revealing good agreement between $\Delta \ln K$ and ΔP_{lrc}.[28] The largest increase of ΔP_{lrc} was observed in the metal-binding region between residues 37 to 40. In this region, the amino acids are conserved so that the higher rigidity of the hyperthermophilic protein is not determined by the amino acid composition. Rather, additional hydrophobic contacts with surrounding residues are the cause. When adding these contacts to the mesophilic rubredoxin, increases in the protein's rigidity and melting temperature were predicted. Thus, the approach seems promising for investigating structural determinants of thermostability differences between homologous proteins. However, as both the mesophilic and thermophilic structures are required for calculating ΔP_{lrc}, the approach seems less applicable for a prospective study where amino acids need to be predicted in a mesophilic structure that, when mutated, may lead to an increase in a protein's thermal stability.

With the aim to relate protein stability to conformational flexibility, Jacobs et al.[97] developed the distance constraint model (DCM). The DCM is based on a rigorous free energy decomposition scheme representing a native protein structure in terms of fluctuating constraint networks.[98] Network rigidity is calculated using the FIRST approach to model enthalpy–entropy compensation in a way that resolves the problem of nonadditivity of component entropies. DCM allows us to calculate quantitative stability/flexibility relationships (QSFR)[99] based on a variety of QSFR descriptors: (1) global stability as a function of a flexibility order parameter θ, defined as the average number of independent, disordered torsion constraints divided by the total number of protein residues; (2) local flexibility profiles, which are based on the probability that a backbone dihedral angle can rotate P_R; and (3) cooperativity correlation plots, which identify regions that are either rigidly correlated, flexibly correlated, or not correlated. Initially, Livesay and Jacobs[99] investigated the stability–flexibility relationship of a pair of orthologous mesophilic and thermophilic RNAse H proteins by DCM. The authors found that the thermophilic protein is more stable than its mesophilic counterpart at any given temperature, but otherwise the global stability profiles are markedly similar for the two proteins at appropriately shifted temperatures. A similar result has been obtained for profiles of the rigidity order parameter and cluster configuration entropy of TLP and IPMDH, respectively, as computed by CNA.[27] Likewise, local flexibility profiles based on P_R were found to be conserved at respective melting temperatures, as were correlated conformational changes. The latter suggests that these changes are necessary for functional efficiency. The same approach was used by Mottonen et al.[100] to investigate thermodynamic properties of nine oxidized and three reduced members of the thioredoxin family. Again, backbone flexibility was found to be well conserved across the family, whereas cooperativity correlation describing mechanical and thermodynamic couplings between residue pairs exhibit distinct features. The latter could be explained in that small-scale structural variations are amplified into discernible global differences by propagating mechanical couplings through a hydrogen bond network.

The DCM is remarkable in that it allows the calculation of thermodynamic quantities and average mechanical properties from the ensemble of constraint networks it produces. However, for this to work it requires the fitting of three parameters to protein-specific experimental thermodynamic data, such as excess heat capacity, which is not available in all cases. As the parameters are only transferable within a protein family,[100] this is expected to limit the applicability of the DCM. As for investigating thermostability, the DCM has been applied so far to only two protein families that are limited in size (less than 165 residues) and show a clear two-state folding behavior. Also, no attempt has been made to predict sites that should be mutated in order to increase a protein's thermostability.

3.5 CONCLUSION

We described promising computational approaches based on rigidity theory that allow researchers to investigate and improve the thermal adaptation of proteins. These approaches are based on constraint network representations of native protein structures and thus use simplified interaction models compared with traditional force fields, with the benefit of a much-increased computational efficiency compared to MD simulations. The methods provide sound insights into how structural differences between homologous proteins lead to discernable global differences in terms of thermostability. At the same time, they demonstrate that enzyme function requires certain regions of a protein to be flexible at the working temperature of an enzyme and allow us to relate changes in the flexibility of those regions, induced either by a temperature change or a mutation, to experimentally observed losses of the enzymes' functions. One of the methods, CNA, already proved successful in improving the thermal stability of a phytase in a prospective study based on predicted structural "weak spots" that were subsequently stabilized by mutations.

Despite these successes, several open points remain. These include (1) improving the interaction model (e.g., by considering the influence of the location of a hydrogen bond on its stability), (2) reducing the sensitivity of the rigidity analysis with respect to structural deviations (e.g., by using ensembles of input structures), (3) developing a quantitative understanding of the influence of flexibility changes on changes in enzyme activity, and (4) by developing an efficient method for testing the actual influence of a mutant on protein stability and enzyme activity.

ACKNOWLEDGMENTS

We are grateful to Simone Fulle and Prakash Rathi (Heinrich-Heine-University, Düsseldorf) for critically reading the manuscript. Part of this work was supported by BASF SE, Ludwigshafen.

REFERENCES

1. Vogt, G., and P. Argos. 1997. Protein thermal stability: Hydrogen bonds or internal packing? *Fold Des* 2: 40–46.

2. Kauzmann, W. 1959. Some factors in the interpretation of protein denaturation. *Adv Protein Chem* 14: 1–63.
3. Dill, K. A. 1990. Dominant forces in protein folding, *Biochemistry (Mosc)* 29: 7133–55.
4. Jaenicke, R., and G. Böhm. 1998. The stability of proteins in extreme environments. *Curr Opin Struct Biol* 8: 738–48.
5. Sterner, R., and W. Liebl. 2001. Thermophilic adaptation of proteins. *Crit Rev Biochem Mol Biol* 36: 39–106.
6. Li, W. F., X. X. Zhou, and P. Lu. 2005. Structural features of thermozymes. *Biotechnol Adv* 23: 271–81.
7. Vieille, C., and G. J. Zeikus. 2001. Hyperthermophilic enzymes: Sources, uses, and molecular mechanisms for thermostability. *Microbiol Mol Biol Rev* 65: 1–43.
8. Zhou, X. X., Y. B. Wang, Y. J. Pan, and W. F. Li. 2008. Differences in amino acids composition and coupling patterns between mesophilic and thermophilic proteins. *Amino Acids* 34: 25–33.
9. Liu, J., H. M. Yu, and Z. Y. Shen. 2008. Insights into thermal stability of thermophilic nitrile hydratases by molecular dynamics simulation. *J Mol Graph Model* 27: 529–35.
10. Robinson-Rechavi, M., A. Alibes, and A. Godzik. 2006. Contribution of electrostatic interactions, compactness and quaternary structure to protein thermostability: Lessons from structural genomics of *Thermotoga maritima*. *J Mol Biol* 356: 547–57.
11. Szilágyi, A., and P. Závodszky. 2000. Structural differences between mesophilic, moderately thermophilic and extremely thermophilic protein subunits: Results of a comprehensive survey. *Structure* 8: 493–504.
12. Vogt, G., S. Woell, and P. Argos. 1997. Protein thermal stability, hydrogen bonds, and ion pairs. *J Mol Biol* 269: 631–43.
13. Russell, R. J., and G. L. Taylor. 1995. Engineering thermostability: Lessons from thermophilic proteins. *Curr Opin Biotechnol* 6: 370–74.
14. Querol, E., J. A. Perez-Pons, and A. Mozo-Villarias. 1996. Analysis of protein conformational characteristics related to thermostability. *Protein Eng* 9: 265–71.
15. Vieille, C., and J. G. Zeikus. 1996. Thermozymes: Identifying molecular determinants of protein structural and functional stability. *Trends Biotechnol* 14: 183–90.
16. Jaenicke, R., and G. Böhm. 1998. The stability of proteins in extreme environments. *Curr Opin Struct Biol* 8: 738–48.
17. Jaenicke, R. 1991. Protein stability and molecular adaptation to extreme conditions. *Eur J Biochem* 202: 715–28.
18. Laman, G. 1970. Graphs and rigidity of plane skeletal structures. *J Eng Math* 4: 331–40.
19. Roth, B. 1981. Rigid and flexible frameworks. *Am Math Monthly* 88: 6–21.
20. Thorpe, M. F. 1995. Bulk and surface floppy modes. *J Non-Cryst Solids* 182: 135–42.
21. Maxwell, J. C. 1864. On the calculation of the equilibrium and stiffness of frames. *Philos Mag* 27: 294–99.
22. Crapo, H. 1979. Structural rigidity. *Struct Topol* 1: 26–45.
23. Roth, B., and W. Whiteley. 1981. Tensegrity frameworks. *Trans Am Math Soc* 265: 419–46.
24. Jacobs, D. J., A. J. Rader, L. A. Kuhn, and M. F. Thorpe. 2001. Protein flexibility predictions using graph theory. *Proteins* 44: 150–65.
25. Hespenheide, B. M., A. J. Rader, M. F. Thorpe, and L. A. Kuhn. 2002. Identifying protein folding cores from the evolution of flexible regions during unfolding. *J Mol Graph Model* 21: 195–207.
26. Jacobs, D. J., D. R. Livesay, J. Hules, and M. L. Tasayco. 2006. Elucidating quantitative stability/flexibility relationships within thioredoxin and its fragments using a distance constraint model. *J Mol Biol* 358: 882–904.

27. Radestock, S., and H. Gohlke. 2008. Exploiting the link between protein rigidity and thermostability for data-driven protein engineering. *Eng Life Sci* 8: 507–22.

28. Rader, A. J. 2009. Thermostability in rubredoxin and its relationship to mechanical rigidity. *Phys Biol* 7: 16002.

29. Karplus, M., and J. A. McCammon. 2002. Molecular dynamics simulations of biomolecules. *Nat Struct Biol* 9: 646–52.

30. Case, D. A. 1994. Normal-mode analysis of protein dynamics. *Curr Opin Struct Biol* 4: 285–90.

31. van Gunsteren, W. F., P. H. Hünenberger, H. Kovacs, A. E. Mark, and C. A. Schiffer. 1995. Investigation of protein unfolding and stability by computer simulation. *Philos Trans R Soc Lond B Biol Sci* 348: 49–59.

32. Sham, Y. Y., B. Y. Ma, C. J. Tsai, and R. Nussinov. 2002. Thermal unfolding molecular dynamics simulation of *Escherichia coli* dihydrofolate reductase: Thermal stability of protein domains and unfolding pathway. *Proteins* 46: 308–20.

33. Pang, J. Y., and R. K. Allemann. 2007. Molecular dynamics simulation of thermal unfolding of *Thermatoga maritima* DHFR. *Phys Chem Chem Phys* 9: 711–18.

34. Purmonen, M., J. Valjakka, K. Takkinen, T. Laitinen, and J. Rouvinen. 2007. Molecular dynamics studies on the thermostability of family 11 xylanases. *Protein Eng Des Sel* 20: 551–59.

35. Creveld, L. D., A. Amadei, R. C. van Schaik, H. A. Pepermans, J. de Vlieg, and H. J. Berendsen. 1998. Identification of functional and unfolding motions of cutinase as obtained from molecular dynamics computer simulations. *Proteins* 33: 253–64.

36. Burioni, R., D. Cassi, F. Cecconi, and A. Vulpiani. 2004. Topological thermal instability and length of proteins. *Proteins* 55: 529–35.

37. Lagrange, J. L. 1788. *Mécanique Analytique*. Paris: Dessaint.

38. Tay, T.-S., and W. Whiteley. 1984. Recent advances in the generic rigidity of structures. *Struct Topol* 9: 31–38.

39. Katoh, N., and S.-I. Tanigawa. 2009. A proof of the molecular conjecture. In *Proceedings of the 25th Annual Symposium on Computational Geometry*, ed. J. Hershberger and E. Fogel, 296–305. New York: ACM Press.

40. Jacobs, D. J., and M. F. Thorpe. 1995. Generic rigidity percolation: The pebble game. *Phys Rev Lett* 75: 4051–54.

41. Jacobs, D. J. 1998. Generic rigidity in three-dimensional bond-bending networks. *J Phys A Math Gen* 31: 6653–68.

42. Thorpe, M. F. 1983. Continuous deformations in random networks. *J Non-Cryst Solids* 57: 355–70.

43. Sartbaeva, A., S. A. Wells, M. M. Treacy, and M. F. Thorpe. 2006. The flexibility window in zeolites. *Nat Mater* 5: 962–65.

44. Rader, A. J., B. M. Hespenheide, L. A. Kuhn, and M. F. Thorpe. 2002. Protein unfolding: Rigidity lost. *Proc Natl Acad Sci USA* 99: 3540–45.

45. Gohlke, H., L. A. Kuhn, and D. A. Case. 2004. Change in protein flexibility upon complex formation: Analysis of Ras-Raf using molecular dynamics and a molecular framework approach. *Proteins* 56: 322–37.

46. Ahmed, A., and H. Gohlke. 2006. Multiscale modeling of macromolecular conformational changes combining concepts from rigidity and elastic network theory. *Proteins* 63: 1038–51.

47. Gohlke, H., and M. F. Thorpe. 2006. A natural coarse graining for simulating large biomolecular motion. *Biophys J* 91: 2115–20.

48. Fulle, S., and H. Gohlke. 2008. Analyzing the flexibility of RNA structures by constraint counting. *Biophys J* 94: 4202–19.

49. Fulle, S., and H. Gohlke. 2009. Constraint counting on RNA structures: Linking flexibility and function. *Methods* 49: 181–88.

50. Dahiyat, B. I., D. B. Gordon, and S. L. Mayo. 1997. Automated design of the surface positions of protein helices. *Protein Sci* 6: 1333–37.
51. Jacobs, D. J., and M. F. Thorpe. 1998. Computer-implemented system for analyzing rigidity of substructures within a macromolecule. U.S. Patent 6014449.
52. Jacobs, D. J., and B. Hendrickson. 1997. An algorithm for two-dimensional rigidity percolation: The pebble game. *J Comput Phys* 137: 346–65.
53. Hespenheide, B. M., D. J. Jacobs, and M. F. Thorpe. 2004. Structural rigidity in the capsid assembly of cowpea chlorotic mottle virus. *J Phys Condens Matter* 16: S5055–64.
54. Wang, Y., A. J. Rader, I. Bahar, and R. L. Jernigan. 2004. Global ribosome motions revealed with elastic network model. *J Struct Biol* 147: 302–14.
55. Fulle, S., and H. Gohlke. 2009. Statics of the ribosomal exit tunnel: Implications for cotranslational peptide folding, elongation regulation, and antibiotics binding. *J Mol Biol* 387: 502–17.
56. Benjwal, S., S. Verma, K. H. Rohm, and O. Gursky. 2006. Monitoring protein aggregation during thermal unfolding in circular dichroism experiments. *Protein Sci* 15: 635–39.
57. Makhatadze, G. I., and P. L. Privalov. 1995. Energetics of protein structure. *Adv Protein Chem* 47: 307–425.
58. Mamonova, T., B. Hespenheide, R. Straub, M. F. Thorpe, and M. Kurnikova. 2005. Protein flexibility using constraints from molecular dynamics simulations. *Phys Biol* 2: S137–47.
59. Stauffer, D., and A. Aharony, eds. 1994. *Introduction to Percolation Theory*, 2nd ed. London: Taylor & Francis.
60. Andraud, C., A. Beghdadi, and J. Lafait. 1994. Entropic analysis of random morphologies. *Physica A* 207: 208–12.
61. Tsang, I. R., and I. J. Tsang. 1999. Cluster size diversity, percolation, and complex systems. *Phys Rev E* 60: 2684–98.
62. Radestock, S. 2009. Entwicklung eines rechnerischen Verfahrens zur Simulation der thermischen Entfaltung von Proteinen und zur Untersuchung ihrer Thermostabilität. Goethe University, Frankfurt am Main.
63. Eijsink, V. G. H., O. R. Veltman, W. Aukema, G. Vriend, and G. Venema. 1995. Structural determinants of the stability of thermolysin-like proteinases. *Nat Struct Biol* 2: 374–79.
64. Gaseidnes, S., B. Synstad, X. H. Jia, H. Kjellesvik, G. Vriend, and V. G. Eijsink. 2003. Stabilization of a chitinase from *Serratia marcescens* by Gly → Ala and Xxx → Pro mutations. *Protein Eng* 16: 841–46.
65. Adekoya, O. A., and I. Sylte. 2009. The thermolysin family (M4) of enzymes: Therapeutic and biotechnological potential. *Chem Biol Drug Des* 73: 7–16.
66. Imanaka, T., M. Shibazaki, and M. Takagi. 1986. A new way of enhancing the thermostability of proteases. *Nature* 324: 695–97.
67. van den Burg, B., H. G. Enequist, M. E. van der Haar, V. G. Eijsink, B. K. Stulp, and G. Venema. 1991. A highly thermostable neutral protease from *Bacillus caldolyticus*: Cloning and expression of the gene in *Bacillus subtilis* and characterization of the gene product. *J Bacteriol* 173: 4107–15.
68. Hardy, F., G. Vriend, O. R. Veltman, B. van der Vinne, G. Venema, and V. G. Eijsink. 1993. Stabilization of *Bacillus stearothermophilus* neutral protease by introduction of prolines. *FEBS Lett* 317: 89–92.
69. van den Burg, B., B. W. Dijkstra, G. Vriend, B. van der Vinne, G. Venema, and V. G. Eijsink. 1994. Protein stabilization by hydrophobic interactions at the surface. *Eur J Biochem* 220: 981–985.
70. Kidokoro, S., Y. Miki, K. Endo, A. Wada, H. Nagao, T. Miyake, A. Aoyama, T. Yoneya, K. Kai, and S. Ooe. 1995. Remarkable activity enhancement of thermolysin mutants. *FEBS Lett* 367: 73–76.

71. Veltman, O. R., G. Vriend, P. J. Middelhoven, B. van den Burg, G. Venema, and V. G. Eijsink. 1996. Analysis of structural determinants of the stability of thermolysin-like proteases by molecular modelling and site-directed mutagenesis. *Protein Eng* 9: 1181–89.

72. van den Burg, B., G. Vriend, O. R. Veltman, G. Venema, and V. G. Eijsink. 1998. Engineering an enzyme to resist boiling. *Proc Natl Acad Sci USA* 95: 2056–60.

73. Kirino, H., M. Aoki, M. Aoshima, Y. Hayashi, M. Ohba, A. Yamagishi, T. Wakagi, and T. Oshima. 1994. Hydrophobic interaction at the subunit interface contributes to the thermostability of 3-isopropylmalate dehydrogenase from an extreme thermophile, *Thermus thermophilus*. *Eur J Biochem* 220: 275–81.

74. Numata, K., M. Muro, N. Akutsu, Y. Nosoh, A. Yamagishi, and T. Oshima. 1995. Thermal stability of chimeric isopropylmalate dehydrogenase genes constructed from a thermophile and a mesophile. *Protein Eng* 8: 39–43.

75. Sakurai, M., H. Moriyama, K. Onodera, S. Kadono, K. Numata, Y. Hayashi, J. Kawaguchi, A. Yamagishi, T. Oshima, and N. Tanaka. 1995. The crystal-structure of thermostable mutants of chimeric 3-isopropylmalate dehydrogenase, 2t2m6t. *Protein Eng* 8: 763–67.

76. Kotsuka, T., S. Akanuma, M. Tomuro, A. Yamagishi, and T. Oshima. 1996. Further stabilization of 3-isopropylmalate dehydrogenase of an extreme thermophile, *Thermus thermophilus*, by a suppressor mutation method. *J Bacteriol* 178: 723–27.

77. Aoshima, M., and T. Oshima. 1997. Stabilization of *Escherichia coli* isopropylmalate dehydrogenase by single amino acid substitution. *Protein Eng* 10: 249–54.

78. Akanuma, S., A. Yamagishi, N. Tanaka, and T. Oshima. 1998. Serial increase in the thermal stability of 3-isopropylmalate dehydrogenase from *Bacillus subtilis* by experimental evolution. *Protein Sci* 7: 698–705.

79. Numata, K., Y. Hayashi-Iwasaki, K. Yutani, and T. Oshima. 1999. Studies on interdomain interaction of 3-isopropylmalate dehydrogenase from an extreme thermophile, *Thermus thermophilus*, by constructing chimeric enzymes. *Extremophiles* 3: 259–62.

80. Hori, T., H. Moriyama, J. Kawaguchi, Y. Hayashi-Iwasaki, T. Oshima, and N. Tanaka. 2000. The initial step of the thermal unfolding of 3-isopropylmalate dehydrogenase detected by the temperature-jump Laue method. *Protein Eng* 13: 527–33.

81. Nemeth, A., A. Svingor, M. Pöcsik, J. Dobó, C. Magyar, A. Szilágyi, P. Gál, and P. Závodszky. 2000. Mirror image mutations reveal the significance of an intersubunit ion cluster in the stability of 3-isopropylmalate dehydrogenase. *FEBS Lett* 468: 48–52.

82. Numata, K., Y. Hayashi-Iwasaki, J. Kawaguchi, M. Sakurai, H. Moriyama, N. Tanaka, and T. Oshima. 2001. Thermostabilization of a chimeric enzyme by residue substitutions: Four amino acid residues in loop regions are responsible for the thermostability of *Thermus thermophilus* isopropylmalate dehydrogenase. *Biochim Biophys Acta* 1545: 174–83.

83. Tamakoshi, M., Y. Nakano, S. Kakizawa, A. Yamagishi, and T. Oshima. 2001. Selection of stabilized 3-isopropylmalate dehydrogenase of *Saccharomyces cerevisiae* using the host-vector system of an extreme thermophile, *Thermus thermophilus*. *Extremophiles* 5: 17–22.

84. Ohkuri, T., and A. Yamagishi. 2003. Increased thermal stability against irreversible inactivation of 3-isopropylmalate dehydrogenase induced by decreased van der Waals volume at the subunit interface. *Protein Eng* 16: 615–21.

85. Shimada, J., A. V. Ishchenko, and E. I. Shakhnovich. 2000. Analysis of knowledge-based protein-ligand potentials using a self-consistent method. *Protein Sci* 9: 765–75.

86. Tamakoshi, M., A. Yamagishi, and T. Oshima. 1995. Screening of stable proteins in an extreme thermophile, *Thermus thermophilus*. *Mol Microbiol* 16: 1031–36.

87. Akanuma, S., C. X. Qu, A. Yamagishi, N. Tanaka, and T. Oshima. 1997. Effect of polar side chains at position 172 on thermal stability of 3-isopropylmalate dehydrogenase from *Thermus thermophilus*. *FEBS Lett* 410: 141–44.

88. Qu, C., A. Akanuma, H. Moriyama, N. Tanaka, and T. Oshima. 1997. A mutation at the interface between domains causes rearrangement of domains in 3-isopropylmalate dehydrogenase. *Protein Eng* 10: 45–52.

89. Akanuma, S., A. Yamagishi, N. Tanaka, and T. Oshima. 1996. Spontaneous tandem sequence duplications reverse the thermal stability of carboxyl-terminal modified 3-iso-propylmalate dehydrogenase. *J Bacteriol* 178: 6300–304.

90. Nurachman, Z., S. Akanuma, T. Sato, T. Oshima, and N. Tanaka. 2000. Crystal structures of 3-isopropylmalate dehydrogenases with mutations at the C-terminus: Crystallographic analyses of structure-stability relationships. *Protein Eng* 13: 253–58.

91. Haefner, S., A. Knietsch, E. Scholten, J. Braun, M. Lohscheidt, and O. Zelder. 2005. Biotechnological production and applications of phytases. *Appl Microbiol Biotechnol* 68: 588–97.

92. Lehmann, M., C. Loch, A. Middendorf, D. Studer, S. F. Lassen, L. Pasamontes, A. P. G. M. van Loon, and M. Wyss. 2002. The consensus concept for thermostability engineering of proteins: Further proof of concept. *Protein Eng* 15: 403–11.

93. Somero, G. N. 1978. Temperature adaptation of enzymes: Biological optimization through structure-function compromises. *Annu Rev Ecol Syst* 9: 1–29.

94. Radestock, S., and H. Gohlke. 2011. Protein rigidity and thermophilic adaptation. *Proteins* 79: 1089–108.

95. de Kreij, A., B. van den Burg, G. Venema, G. Vriend, V. G. Eijsink, and J. E. Nielsen. 2002. The effects of modifying the surface charge on the catalytic activity of a thermo-lysin-like protease. *J Biol Chem* 277: 15432–38.

96. Yasukawa, K., and K. Inouye. 2007. Improving the activity and stability of thermolysin by site-directed mutagenesis. *Biochim Biophys Acta* 1774: 1281–88.

97. Jacobs, D. J., S. Dallakyan, G. G. Wood, and A. Heckathorne. 2003. Network rigidity at finite temperature: Relationships between thermodynamic stability, the nonadditivity of entropy, and cooperativity in molecular systems. *Phys Rev E Stat Nonlin Soft Matter Phys* 68: 061109.

98. Livesay, D. R., S. Dallakyan, G. G. Wood, and D. J. Jacobs. 2004. A flexible approach for understanding protein stability. *FEBS Lett* 576: 468–76.

99. Livesay, D. R., and D. J. Jacobs. 2006. Conserved quantitative stability/flexibility relationships (QSFR) in an orthologous RNase H pair. *Proteins* 62: 130–43.

100. Mottonen, J. M., M. Xu, D. J. Jacobs, and D. R. Livesay. 2009. Unifying mechanical and thermodynamic descriptions across the thioredoxin protein family. *Proteins* 75: 610–27.

4 Thermostable Subtilases (Subtilisin-Like Serine Proteinases)

Magnús M. Kristjánsson

CONTENTS

4.1 INTRODUCTION

In the MEROPS peptidase database (http://merops.sanger.ac.uk/),[1] subtilases, or subtilisin-like serine proteinases, are classified as the superfamily, or clan, SB of the serine proteinases. They are a highly diverse group of proteases, occurring in viruses, archaea, and eubacteria among prokaryotes, fungi, yeasts, as well as in higher eukaryotes, including plants and animals.[2,3] They are classified further in the MEROPS database to the S8 peptidase family (subtilisins), which is divided into two subfamilies: S8A, typified by the subtilisins, and S8B, of which the yeast enzyme kexin is an example (this enzyme was the first eukaryotic subtilase to be identified).[4,5] The active sites of the subtilases are characterized by a catalytic triad of Asp-His-Ser (DHS) residues, which they share with proteases of the chymotrypsin-like superfamily (SA), however, these catalytic residues occur in different order within the structures of members of the two superfamilies. In the subtilase superfamily the order is DHS, but it is HDS in the chymotrypsin-like enzymes. The subtilases have also been subdivided into six families, based on their sequence homology, where each family is named after a well-characterized representative enzyme belonging to the family (i.e., the subtilisin, thermitase, proteinase K, lantibiotic peptidase, kexin, and pyrolysin families).[2]

While the "classic" Asp-His-Ser catalytic triad of the S8 family is the most widespread in nature, a variety of catalytic triad configurations are observed.[3,6] In fact, extreme sequence variability exists among subtilases, found to extend to two of the three catalytic residues, with the nucleophilic Ser residue invariably being found in all subtilases.[3] Among members of the related sedolisin family (S53), the catalytic triad is Glu-Asp-Ser, where Glu replaces His as a general acid base of the classic triad. Furthermore, in the sedolisins, an Asp residue replaces Asn in the oxyanion hole of subtilases and the enzymes have acidic pH optima, but protein folds are clearly related to the subtilisins, but not to the pepstatin-sensitive apartic proteinases (http://merops.sanger.ac.uk/). The different variations on the catalytic triads of the serine proteinases have been reviewed[6] and the implications for their classifications have been discussed.[3] An extensive database, the Prokaryotic Subtilase Database (www.cmbi.ru.nl./subtilases), provides access to all information available on identified subtilases in prokaryotes, which are subdivided into families based on conserved sequences around these variable catalytic triads.[3]

Subtilases have highly diverse functions in living organisms. In prokaryotes they are most often extracellular, secreted outside the cell, and play roles in nutrition, by digestion of protein substrates providing peptides and amino acids for growth, or for host invasion by pathogens.[3] Subtilases have also been shown to be involved in various intra- and extracellular precursor activation and maturation reactions in both prokaryotes and eukaryotes.[3,7,8] These reactions may involve maturation of the α mating factor; killer toxin in yeast, as in the case of kexin;[4] or limited proteolysis of proproteins in the activation of growth factors, hormones, enzymes, or receptors, as for mammalian proprotein convertases.[9]

Apart from their various physiological functions, subtilases are highly important in technical enzyme applications, of which their use in detergents is by far the most prominent with respect to market volume and tonnage.[10] Among these, subtilisins from *Bacillus* species have been applied to the greatest extent and provide all the proteases for the detergent industry, which amounts to a production of hundreds of tons of pure enzyme each year.[10] These proteases have alkaline activity optima (pH 9–11) and are therefore appropriate for digesting proteinaceous stains on fabrics in highly alkaline detergent solutions.[10,11] These enzymes are generally quite stable at moderately high temperatures (60°C to 65°C).[12] For the past two decades considerable research has gone into the search for detergent proteases with improved properties, both with respect to activity and stability under the demanding conditions used in washing, including high temperatures and strongly oxidizing or chelating conditions. A number of subtilases from thermophiles and hyperthermophiles have been characterized, and protein engineering approaches, based on both rational design using site-directed mutagenesis and directed evolution, as well as random mutagenesis approaches, have been exploited in an effort to improve the properties of detergent subtilases. Despite these efforts, however, fewer than 15 different subtilases or variants, all originating from *Bacilli*, are used in detergents worldwide.[10] A number of novel subtilases of different origins (e.g., from extremophiles) are described in the scientific literature each year, and subtilases are still extensively studied by protein engineering approaches. Some of these wild-type or mutant enzymes would be expected to have improved properties for detergent applications but have not been used for that purpose. A major reason may be that changing the production lines for the present detergent proteases for subtilases with improved properties may not be cost effective.[12] Partly for the reasons mentioned (i.e., the industrial importance), subtilases of different origins have been extensively studied from both structural and functional standpoints, and few enzymes, if any, have been as extensively studied by protein engineering as the subtilisins.[10,13–20] This chapter will attempt to summarize and reflect on the present state of knowledge on the structural stability of this group of enzymes. It will focus on the subtilase fold of enzymes of (hyper)thermophilic origin and stabilized mutants that have provided insight into the structural basis for stabilization of this protein fold. The prokaryotic subtilases have been studied to the greatest extent with respect to both structure and function, and the discussion will focus on these well-characterized enzymes.

4.2 STRUCTURE AND FOLDING OF THE SUBTILASES

4.2.1 STRUCTURE

As mentioned, the subtilases are a highly diverse group of serine proteinases. They are most often produced as multidomain proteins, with considerable variability in types, numbers, and different combinations/organizations of the domains or modules involved. There are currently 275 registered architectures in the protein family database (pfam), representing different combinations of domains of the S8 peptidase or subtilase family (http://pfam.sanger.ac.uk//family/PF00082). These cover 6532 sequences from 1079 species and currently (February 2011) there are 242 solved three-dimensional (3D) structures of subtilases deposited in the Protein Data Bank (PDB) representing 32 different serine proteases containing the subtilase domain structure (Figure 4.1).

The subtilisin-like fold belongs to the class of α/β proteins composed mainly of parallel β sheets arranged in $\beta\alpha\beta$ units, according to the SCOP (http://scop.mrc-lmb.cam.ac.uk/scop/) and CATH (http://www.cathdb.info/) databases. The fold is characterized by a three-layer ($\alpha\beta\alpha$) sandwich with a parallel β sheet of seven strands aligned in the order 2314567 and with a rare left-handed crossover connection between strands 2 and 3 (SCOP) (Figure 4.1). In the CATH database, it is designed in the topology of the Rossmann fold. The α helices (seven in VPR in Figure 4.1) surround the β sheet and are connected to the strands by loops that lie on the surface of the protein. In addition, there is generally an antiparallel β sheet close to the C-terminus of the proteins. The active site is arranged within a cleft at the surface of the molecule, with the catalytic Ser (S220) and His (H57) located at the ends of adjacent α helices (B and F in Figure 4.1a), and with Asp (D37) at a β strand of the central β sheet.[22] The substrate-binding site is located within this cleft and the N-terminal part of a substrate backbone binds as a central strand in the antiparallel β sheet between residues 100 to 102 on one side of the cleft and 125 to 127 on the other.[22] As will be discussed in more detail later, different subtilases have different numbers of calcium-binding sites and disulfide bonds within their subtilase domain that contribute to their stability.

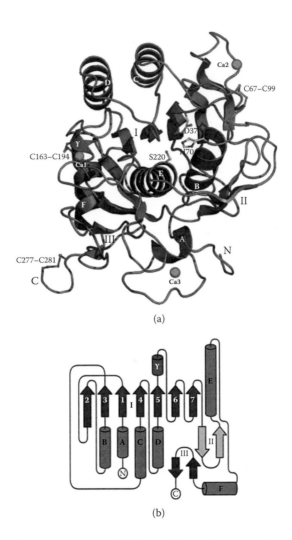

(a)

(b)

FIGURE 4.1 (a) Crystal structure of VPR, a subtilase from a psychrotrophic *Vibrio* sp. PA-44 (PDB entry 1sh7). The residues of the catalytic triad are shown in yellow, calcium ions are shown as green spheres, and disulfide bonds are in orange. (b) A topology diagram of the VPR structure. (From J. Arnórsdóttir, M. M. Kristjánsson, and R. Ficner, Crystal Structure of a Subtilisin-Like Serine Proteinase from a Psychrotrophic *Vibrio* Species Reveals Structural Aspects of Cold Adaptation, 2005, *FEBS Journal* 272: 832–845. Copyright Wiley-VCH Verlag GmbH & Co. KGaA. Reproduced with permission.) **(See color insert.)**

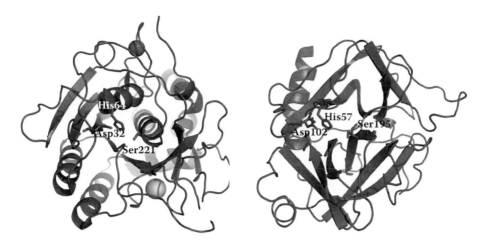

FIGURE 4.2 Schematic representations of typical 3D structures of serine proteases belonging to the subtilase (gray) and chymotrypsin-like (blue) superfamilies. The structures are those of subtilisin Carlsberg (PBD entry 1scn) and bovine chymotrypsin (PDB entry 2gmt). Labeled are residues of the catalytic triad, following the order DHS for subtilases and HDS for the chymotrypsin-like enzyme. Bound calcium ion to subtilisin is shown as a red sphere and natrium ion is in yellow. **(See color insert.)**

The subtilases have been extensively studied structurally, with several high-resolution 3D protein structures containing the subtilase domain in the PDB, some of which are solved at atomic resolution.[23–26] Despite a low sequence homology, the proteases having this fold share a high structural similarity.[3] However, while they share the same active site, the catalytic triad, and the same mechanism of action in cleaving peptide bonds, the subtilases share no structural similarity whatsoever with proteases belonging to the chymotrypsin-like protease superfamily, which are characterized by β barrel structures (Figure 4.2).

All prokaryotic subtilases are synthesized as precursor proteins (i.e., as preproenzymes), which consist of a signal peptide required for secretion of the proenzyme across a membrane, and a prodomain, essential for correct folding of the subtilase domain.[27–31] The prodomain thus has a chaperone-like function and is often referred to as an intramolecular chaperone (IMC).[29–36] In the case of subtilisins E and BPN′, the signal peptide is 29 amino acid residues, the IMC prodomain is 77 residues, and the active subtilase domain is 275 residues.[27–31] Several subtilases also have C-terminal propeptides, as in the case of the thermostable aqualysin I (AQUI) from *Thermus aquaticus*[37] and the cold-adapted subtilase VPR from *Vibrio* sp. PA44.[38] Both these proteinases, which belong to the proteinase K family of subtilases, rely on greater than 100 residue long N- and C-terminal prodomains for processing of active extracellular proteases. As for the subtilisins, the N-terminal prodomain is an intramolecular chaperone, whereas the C-terminal propeptide facilitates the extracellular secretion of the enzyme[39,40] (Figure 4.3). Despite the difference in size and the fact that sequence identity between the 77 residue and 113 residue prodomains of subtilisin E and AQUI, respectively, is only 21%, the prodomains are clearly related both functionally and structurally.[35] This is supported by experimental findings showing

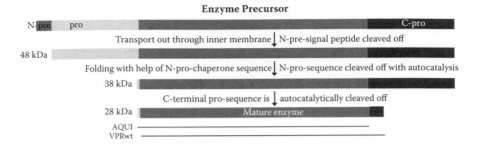

FIGURE 4.3 Processing of precursor proteins of VPR, a psychrotrophic subtilase from *Vibrio* sp. PA-44, and AQUI from the thermophile *Thermus aquaticus* YT-1 upon folding and secretion. A signal peptide is shown in red, the N-terminal prodomain is in yellow, the C-terminal domain is in green, and the mature subtilase domain is in gray. The mature AQUI is 2 residues longer at the N-terminus and 15 residues shorter at the C-terminus than the wild-type VPR (VPRwt). **(See color insert.)**

that the prodomain of AQUI (ProA) is capable of chaperoning subtilisin (BPN′) when added in *trans*, as well as to a large extent in the intramolecular (*cis*) refolding of denatured subtilisin E in a chimera of the enzyme containing ProA as the IMC.[41] Furthermore, the ProA domain was a potent inhibitor of subtilisin BPN′.[35]

4.2.2 Folding

Considerable research efforts have gone into studying both the mechanistic[28–31,33,34,41–50] and structural[31,48,51–55] aspects of the folding pathway for subtilisins and the role of the prodomain in the folding process. The presence of the prodomain is essential for correct folding of the subtilase domain, but must be released upon folding to generate the active protease. The maturation of prosubtilisin into enzymatically active subtilisin involves at least three distinct steps: (1) folding of the precursor that is mediated by the IMC prodomain; (2) autoproteolytic cleavage of the covalent linkage between the IMC domain and the mature protein that results in a noncovalently linked IMC, subtilisin stoichiometric complex; and (3) autodegradation of the IMC domain to give the mature and active subtilisin.[25,36,53,56,57] While the process described is an intramolecular process, it is clear that the 77 residue prodomain of the subtilisins will catalyze folding of the active protease even as a separate polypeptide chain.[31–33,35,41] In the second step of the folding process, that is, in the noncovalent IMC:subtilisin complex, the prodomain is bound as a competitive inhibitor. In the folded state the propeptide and subtilisin form a tight complex with a K_a of 2×10^8/M at 25°C.[31,44] By itself, in an unbound state, the propeptide/domain is largely unstructured and is 97% unfolded, even under optimal folding conditions corresponding to only about −2 kcal/mol at 25°C in terms of the free energy of unfolding.[31,46] In the complex with subtilisin, however, the 77 residue prodomain folds into a compact stable structure comprising a four-stranded antiparallel β sheet and two three-turn helices[46,52] (Figure 4.4). For active, mature subtilisin to form, the prodomain is released from the complex and, due to its marginal stability, it unfolds and is readily autocatalytically degraded by the active proteinase.[34–36] This autocatalytic degradation is necessary for

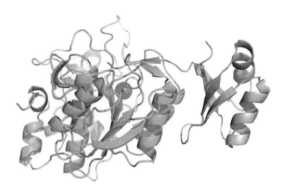

FIGURE 4.4 The structure of prodomain:subtilisin BPN complex (PDB entry 1spb). The subtilisin domain is shown in green and the residue 77 prodomain is shown in brown. (**See color insert.**)

the folded subtilisin to be active as a protease, otherwise the propeptide would inhibit its activity. In the IMC:subtilisin complex, the β sheet of the folded prodomain packs tightly against the two surface helices (helices C and D in Figure 4.1) (Figure 4.4). At the domain interface, acidic residues on the propeptide (e.g., Glu69 and Asp71 in the propeptide:subtilisin E complex) form helix caps with the N-termini of the two helices.[31,53,58,59] There is also a series of hydrophobic residues at this interface site and a network of 27 hydrogen bonds to further stabilize the interface.[31,42,47,53] The C-terminal (residues 72 to 77) extends out of the center of the prodomain and binds in a substrate-like manner at the substrate-binding cleft of the enzyme, with residues 74 to 77 in the prodomain occupying subsites S1 to S4, respectively.[31,42,59] Almost all contacts between the prodomain and subtilisin are made with residues 100 to 144. These residues encompass a 45 residue αβα substructure, including the two helices and the loop-β-loop connecting region between them. This region includes most of the substrate-binding site in the subtilisins. It is believed that this αβα substructure may act as a folding nucleus, with subsequent folding propagating into the N- and C-terminal regions on binding of the IMC prodomain.[31,42,59] In the absence of the IMC domain, refolding of the subtilisin domain results in a stable but inactive intermediate, which has a native-like secondary structure, but little tertiary structure, and has compactness of that between the native and the unfolded states.[29,30] Thus on *in vitro* folding of the protein, it gets trapped in a molten globule-like state that will not fold to its native state because of a high-energy state activation barrrier.[29,30,42,44,58] The presence of the IMC prodomain lowers the activation barrier by binding to the intermediate, as described above, and thereby catalyzes its folding to the native, active enzyme.

Prodomain catalyzed folding appears to be a common characteristic of extracellular proteases, including those of serine proteinases of both superfamilies (i.e., the subtilases and the chymotrypsin-like enzymes).[36,42] Besides convergent evolution among their active site geometries, these enzymes have also coevolved with such N-terminal IMC domains to mediate their folding across a high transition state energy barrier on the folding pathway toward the highly stabilized native structures of those enzymes.[42,58,60,61] For subtilisin BPN′ it has been estimated that the interaction of the prosequence stabilized the transition state for folding in excess of 7.5 kcal/mol, accelerating the folding rate by more than five orders of magnitude.[29,59] The native structure of the enzyme is highly stabilized, as the segregated prodomain is readily cleaved due to its marginal stability and is therefore not present for catalyzing unfolding of the protease back over the high-energy barrier toward the unfolded state. The role of the prodomain thus appears to kinetically trap the protease in an active conformation. Such kinetic stabilization, characterized by very slow unfolding kinetics rather than thermodynamics, undoubtedly plays an important role in the stabilization of extracellular bacterial proteases, which have to be able to withstand harsh environmental conditions. In highly proteolytic environments, both partially and fully unfolded proteins face potential inactivation through degradation or aggregation; hence, slowing unfolding can greatly extend the protein's functional lifetime.[61] Such kinetic stabilization has been extensively studied in the case of the bacterial chymotrypsin-like enzyme α-lytic protease (α-LP).[60,62–67] α-Lytic protease is synthesized with a large 166 amino acid residue prodomain that acts as an IMC, similar to that of the prodomains of the subtilases. In the absence of the prodomain, the enzyme folds to an inactive, molten globule-like intermediate state (I).[60,63] Without the prodomain, its folding to the native state is extremely slow against a high-energy barrier of folding (~30 kcal/mol) with an estimated $t_{1/2}$ of 1800 years at 4°C.[60,63] This high activation barrier is surmounted by the foldase or IMC activity of the prodomain, which accelerates folding by a factor of nearly 10^{10}, or by lowering the energy barrier by 18.2 kcal/mol.[60,63,66] As in the case of subtilisin, the prodomain assists in folding either when supplied in *cis* or in *trans*. Furthermore, it was shown that both the intermediate (I) and the fully unfolded state of α-LP are more stable than the native state (N) by 4 kcal/mol. Thus, the native state is not stabilized thermodynamically, but is instead kinetically stabilized, that is, by being trapped in a metastable native state by a large cooperative energy barrier to unfolding.[60,63] This energy barrier to unfolding was estimated at 26 kcal/mol, leading to very slow unfolding rates with $t_{1/2}$ of ~1.2 years at 4°C.[63,64] The folding free energy diagram for thermodynamic compared with kinetic stability is illustrated in Figure 4.5. Another feature of α-LP associated with the high-energy barrier is the extreme cooperativity of its unfolding transition, for example, in comparison with its thermodynamically more stable homologues such as trypsin or protease B from *Streptomyces griseus* (SGPB).[68] As illustrated in Figure 4.6, the proteolytic resistance, as a measure of cooperativity, is directly correlated with the size of the folding barrier, but inversely correlated with thermodynamic stability.[68] The cooperativity of the unfolding transitions appears to be a critical determinant of protease longevity.[68] Under proteolytic conditions, these more stable homologues are degraded much faster than α-LP because of partial unfolding resulting from a lesser degree of cooperativity. Because of the high degree of cooperativity, such partial unfolding leading to proteolysis is negligible for α-LP, giving it a functional advantage and longevity as a protease in the harsh extracellular environment in which the protease functions.[61,64,68] A thermostable homologue of α-LP, protease A from the thermophile *Thermobifida fusca,* is a case of an even more kinetically thermostable enzyme that attains its remarkable thermostability from extremely slow unfolding kinetics.[69] Structural comparison of the thermophilic enzyme to that of α-LP, and related mesophilic enzymes, has

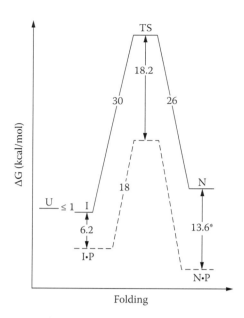

FIGURE 4.5   Free energy diagram of α-LP with and without its prodomain (P) at 4°C. In the absence of P, the unfolded (U) α-LP spontaneously folds to a molten globule-like intermediate (I), which proceeds at an extremely slow rate to the native (N) state through a high-energy folding transition state (TS). The presence of the prodomain provides a catalyzed folding pathway (dashed lines) that lowers the high-energy folding barrier and results in a thermodynamically stable inhibition complex, NP. (From E. L. Cunningham et al., Kinetic Stability as a Mechanism for Protease Longevity, *Proc Natl Acad Sci USA* 96: 11008–11014. Copyright 1999 National Academy of Sciences, USA.)

provided important insight into the structural mechanisms underlying kinetic stability for this group of structurally homologous proteases.[69]

4.3   THERMAL STABILITY OF THE SUBTILASES

Any attempts to rationalize the origins of stability of the subtilases must focus on its dependence on calcium binding. All the prokaryotic subtilases characterized thus far bind calcium ions, which are essential for their stability, and in some cases for activity. The number and location of the calcium-binding sites vary in the different members of the subtilases. For example, subtilisins BPN′ and E have two binding sites[53,70]; proteinase K also has two binding sites, but at different locations.[24,71] Two homologous subtilisins, one from an Antarctic *Bacillus* TA41 and another from the mesophilic *Bacillus sphaericus*, both have five; none, however, that correspond to the high-affinity site found in the other subtilisins.[26] To complicate this picture even further, Tk-subtilisin, a subtilase from the hyperthermophilic archaeon *Thermococcus kodakaraensis*, has seven calcium-binding sites in its subtilase domain, six of which are unique to that enzyme.[48,49] Another subtilase from the *T. kodakaraensis*, Tk-SP, does not contain a calcium site within its subtilase domain, but contains an additional β jelly roll domain that binds two calcium ions and stabilizes the protein at high temperatures.[54,55]

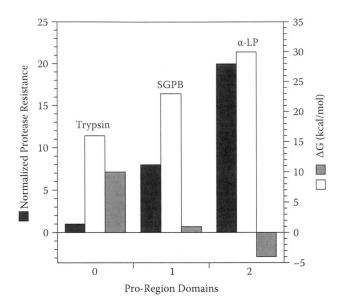

FIGURE 4.6 Comparison of proteolytic resistance (as a measure of cooperativity) and the energetics of folding of the three related chymotrypsin-like proteases: trypsin, SGPB, and α-LP. The black bars represent normalized protease resistance. The increase in the kinetic energy barrier (white bars) and decrease in thermodynamic stability (gray bars) seem to correlate directly with the increase in proteolytic resistance. (From S. M. E. Truhlar, E. L. Cunningham, and D. A. Agard, The Folding Landscape of *Streptomyces griseus* Protease B Reveals the Energetic Costs and Benefits Associated with Evolving Kinetic Stability, *Protein Science*, 2004, 13: 381–390. Copyright Wiley-VCH Verlag GmbH & Co. KGaA. Reproduced with permission.)

It is clear, therefore, that in studies aimed at addressing the stability of subtilases, it is highly critical that all experimental conditions with respect to calcium concentrations and ionic strength be well controlled. It may therefore be helpful to make a distinction between calcium-dependent and calcium-independent stability of subtilases.[16] This can be done by obtaining stability data under defined conditions in the presence of calcium (e.g., 10 mM) or in its absence and in the presence of a chelator, such as ethylenediaminetetraacetic acid (EDTA).[16]

The slow folding rates of subtilases without their prodomains makes it unfeasible to determine the equilibrium constant for folding or unfolding under practical experimental conditions. Estimates indicate that the thermodynamic stability of apo-subtilisin BPN is marginal, hence it is evident that calcium binding makes a dominant contribution to the conformational stability of subtilisins.[16,42,72] This is likely a true statement, at least for most prokaryotic subtilases. Also, as a result of the slow rate of subtilisin (re)folding, most stability measurements are not affected by the equilibrium constant for unfolding ($K_{unfolding} = k_{unfolding}/k_{folding}$), but by the activation energy for unfolding (i.e., the activation barrier for unfolding).[16] Thermal denaturation of subtilases is therefore practically irreversible and is usually determined by following the rates of inactivation under a set of conditions with respect to temperature, pH,

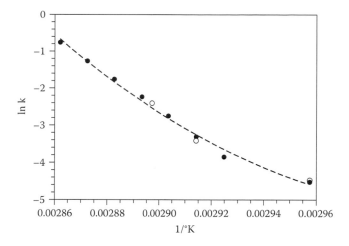

FIGURE 4.7 Comparison of the rates of irreversible thermal inactivation of subtilisin BPN′ with the rate of thermal unfolding measured in 50 mM Tris-HCl, pH 8.0, 50 mM NaCl, 10 mM CaCl$_2$, at temperatures between 65°C and 75°C. Unfolding rates were measured by differential scanning calorimetry. (Reprinted from *Biochimica et Biophysica Acta*, 1543, P. N. Bryan, Protein Engineering of Subtilisin, 203–222. Copyright 2000, with permission from Elsevier).

and solution composition. It is important to control the experimental conditions, as several factors other than thermal denaturation can cause irreversible inactivation by different mechanisms under different conditions (e.g., autolysis, aggregation, or chemical modification). It has been pointed out that studies measuring the rates of inactivation at elevated temperatures are indirectly measuring the rates of unfolding, because unfolding becomes the rate-determining step in irreversible inactivation as the temperature is increased.[16] This has been well demonstrated for subtilisin BPN′, where it has been shown that the rates of unfolding as measured by calorimetry coincide with the rates of thermal inactivation over the temperature range 65°C to 75°C (Figure 4.7).

4.3.1 KINETIC STABILIZATION OF SUBTILASES

Over the last few years, research on the folding mechanisms of proteases such as α-lytic-like proteases[58,61,64,67–69] and subtilisins[42,58,59] have provided some insight into the structural basis for the large kinetic barriers that exist between the unfolded and folded forms of the mature enzymes. For subtilisin BPN′, Fisher et al.[58] studied the various aspects of the bimolecular folding reaction of several specifically designed subtilisin mutants and a stabilized form of the prodomain. They also studied the hydrogen/deuterium exchange reaction by nuclear magnetic resonance (NMR) of subtilisin BPN′ in the free form and in complex with the prodomain to understand how prodomain binding affects the energetics of the folding reaction.[59] Based on their analysis, it was proposed that the inordinately slow folding of subtilisin, and hence the reason for the high kinetic barrier to folding, could be ascribed to accrued effects of two slow and sequential processes: (1) the formation of an unstable and topologically challenged folding intermediate and (2) the proline-limited isomerization of the intermediate to the native state.[58]

The folding reaction for subtilisin can be depicted by a simple scheme (with the prodomain not included):

$$U \leftrightarrow I \leftrightarrow N + metal \leftrightarrow N(metal) \tag{4.1}$$

The folding proceeds through an intermediate (I), which forms early in the process and whose structured regions are slightly nonnative and preferentially stabilized by prodomain binding.[59] Only the native state (N) is stabilized by metal (preferentially calcium) binding, but the two calcium-binding sites are formed late in the folding process, so they contribute little to the stability of the intermediate.[58] However, when formed, calcium binding to the sites greatly stabilizes the native conformation and contributes a significant portion of the high-energy barrier to unfolding.[59] In addition to the instability of the intermediate, its folding to the native form further faces a topological challenge in folding of the $-\beta 2-\alpha C-\beta 3-$ motif with a left-handed crossover connection.[58] Such left-handed $-\beta-\alpha-\beta-$ motifs are very rare in protein structures, but in subtilases it is required to insert β strand 3 between strands 1 and 2, and at the same time to form a proper active site by placing helix C (corresponding to helix B in Figure 4.1), containing His of the catalytic triad, on the same side of the seven-stranded central parallel β sheet as the other Asp and Ser active site residues. Helix C is a principal structural component of this essential left-handed crossover, and interacts with other parts of the protein structure.[58] Calcium-binding site Ca-1 (site A) in subtilisin BPN′ is located at the C-terminal of helix C. It is formed as a 9 residue loop out of the last turn of the helix (Figure 4.8).[42,72] Without calcium bound in the binding loop, the stability of helix C appears to be compromised, as well as its stabilizing interactions to other parts of the protein molecule.[58] Thus, in forming the critical folding intermediate, the challenging left-handed connection would have to be made without the benefits of those stabilizing interactions.[58] Part of the mechanism by which prodomain binding seems to catalyze folding by lowering the energy barrier is by stabilizing this area of α helix C in the essential left-handed crossover between β strands 2 and 3.[59] The critical role of the Ca-1 calcium-binding site in the high-energy barrier to subtilisin folding is shown by the effect of eliminating the calcium loop by deleting residues 75 through 83 from the protein.[43,73] The Δ75-83 mutant of subtilisin was fully active, but with sharply decreased thermostability.[14,44,74] However, the Δ75-83 mutant was capable of independent folding, although the prodomain greatly facilitated its folding.[31] The folding rate was found to be orders of magnitude faster than the wild-type enzyme containing the Ca-1 site.[16,42] The crystal structure of a mutant with the Δ75-83 deletion (Sbt-70) showed that except for the region of the deleted loop, the structure is remarkably similar to the wild-type enzyme, and helix C was extended by about one turn (Figure 4.8).[73]

Based on hydrogen/deuterium exchange rates of amide protons in free and prodomain-complex subtilisin, the binding of the prodomain appears, in addition to the area around helix C and β strands 2 and 3, to facilitate the organization of the entire central β sheet of the protein.[59] In particular, did binding of the prodomain provide

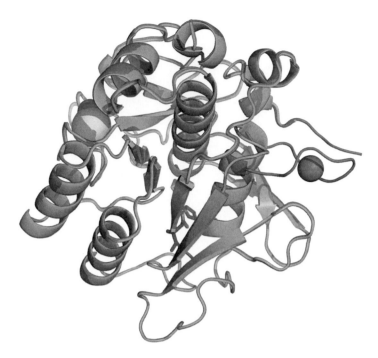

FIGURE 4.8   Superposition of ribbon diagrams of the α-carbon backbone of subtilisin BPN (PDB entry 1sud) (red) and the Δ75-83 mutant (PDB entry 1suc) (green). Calcium ions are shown as red spheres and a potassium ion occupying the Ca-2 site in the Δ75-83 mutant is shown in yellow. Residues of the active site are shown in pink. **(See color insert.)**

a high protection factor in the vicinity of β strands 5, 6, and 7 and the connecting loops between them. These loops provide the ligands for the second calcium-binding site, Ca-2 (site B), in subtilisin. As mentioned earlier, according to Fisher et al.[58] and Sari et al.,[59] the second major contributor to the high kinetic barrier to subtilisin folding is rate-limiting proline isomerization, occurring late in the folding process. Proline *cis* ↔ *trans* isomerization is a rate-limiting step in folding of many proteins.[75,76] There are 14 prolines in subtilisin BPN′, and of those, there are three Pro residues required in their native isomeric form for formation of the Ca-2 site. It was thus suggested that the need for simultaneous rate-limiting isomerization of these prolines into their native conformation in order to create Ca-2 may be a reason for the heightened barrier between I and N in the folding reaction (Equation 4.1).[58] The prodomain appears to stabilize the loops providing the ligands for the Ca-2 site, thus increasing the probability that all three prolines occur in their native isomeric forms.[58] Once the calcium-binding sites Ca-1 and Ca-2 are formed in the transition from I to N, late in the folding reaction, the N state can bind metal (calcium) ions and would be greatly stabilized with respect to I. The binding energy for calcium or other cations to sites 1 and 2 thus contributes significantly to the energy barrier to unfolding of these enzymes, and therefore the stability of the native state. The role of calcium binding in the stabilization of different subtilases is discussed in more detail later.

4.3.2 Thermophilic Subtilases

Subtilases from several thermophilic organisms have been characterized to different extents. Crystal structures are available for some of the thermostable subtilases,[48,55,77–80] and in most cases 3D structures of more thermolabile homologues from meso- and psychrophiles are also available.[21,24–26,71] Such thermostable and thermolabile pairs of homologous proteins are ideal for studying the structural aspects of activity–stability relationships in enzymes (e.g., with respect to temperature adaptation). The following discussion on the structural origins of thermal stability among the subtilases is based on currently available structural and functional data obtained from comparisons of subtilases within the same homology families.[2] For subtilases from the proteinase K family there are high-resolution structures available for the representative enzyme proteinase K from the mesophilic fungi *Tritirachium album* Limber,[24] of VPR from the psychrophile *Vibrio* sp. PA-44,[21] and a crystal structure of AQUI from the thermophile *T. aquaticus* YT1 has also been determined.[80] In the thermitase family, crystal structures for thermitase from the thermophile *Thermoactinomyces vulgaris*[77,78] and a thermostable subtilase from *Bacillus* Ak.1[79] are available. Several mesophilic subtilisins of different *Bacilli* sp. (see, e.g., Jain et al.,[53] Takeuchi et al.,[70] and McPhalan and James[81]) have been extensively studied, but in the subtilisin family there are also structurally well-characterized representatives from psychrophiles, for example, subtilisin S41 from the Antarctic *Bacillus* TA41,[26] and the mesophilic sphericase from *B. sphaericus*.[25,26] To this group also belongs a homologous subtilase from the thermophile *Bacillus* WF146.[82,83] Recently, detailed studies on the crystal structures of two subtilases from the hyperthermophilic archaeon *T. kodakaraensis*, Tk-subtilisin and Tk-SP, in both mature and pro forms, have been published.[48–50,54,55] Furthermore, homology models of subtilases from two thermophilic archaea, *Pyrococcus furiosus* (pyrolysin) and *Thermococcus stetteri* (stetterolysin), have been published.[84]

Summarized below are some of the findings that have been revealed by comparative studies of these enzymes regarding the molecular basis of their thermostability and their function at elevated temperatures.

4.3.2.1 Effect of Calcium Binding on Thermostability

As discussed above, the two calcium-binding sites are formed late in the folding process of the subtilisins and contribute significantly to the energy barrier of unfolding by stabilizing the native state of the enzymes.[16,42,58] Calcium has been implicated in playing a key role in the stability of most subtilases characterized so far. This has been shown for several subtilases both from bacterial[72,79,85,86] as well as eukaryotic organisms, including mammals.[87] To our knowledge, the first reported case of a subtilase structure that did not bind Ca^{2+} in its native state and apparently is independent of Ca^{2+} with respect to both activity and thermostability is subtilase 3 (SBT3) from tomato.[5] Although the general architecture of the calcium-binding regions in the tomato subtilase is similar to that of the calcium sites Ca-1 and Ca-2 in the subtilisins, a structural alternative to calcium binding is realized for the stabilization of this plant subtilase.[5]

For a hyperthermostable subtilase, Tk-SP, from the archaeon *T. kodakaraensis*, none of the known calcium-binding sites of bacterial subtilisins, nor of Tk-subtilisin

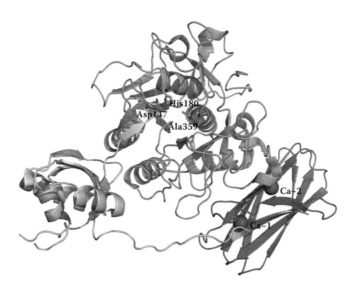

FIGURE 4.9 Backbone structure of ProN-Tk-SP (ProN-TK-S359A) from *T. kodakaraensis* (PDB entry 3afg). The N-prodomain (Lys4-Ala113), subtilase domain (Val114-Tyr421), and C-terminal β jelly roll domain (Ala442-Pro552) are shown in yellow, green, and blue, respectively. The catalytic triad residues Asp147, His180, and Ala359 (active site Ser359Ala mutant) are indicated and the two calcium ions (Ca-1 and Ca-2) of the C-domain are shown as red spheres. **(See color insert.)**

from the same organism, are conserved in its subtilase domain.[54,55] The Tk-SP protease contains an additional domain at the C-terminus of the subtilase domain, characterized by a β jelly roll structural motif, which binds two Ca^{2+} ions (Figure 4.9).[55] Similar β jelly roll-like domains have been described in kexin-like eukaryotic subtilases,[88–90] as well as bacterial subtilases (KP-43) from a *Bacillus* sp. (KSM-KP43)[91] and a kexin-like enzyme from the anaerobe *Aeromonas sobria*.[92] The β barrel domain (the so-called P domains) in the kexin-like subtilases have been implicated in the folding and thermodynamic stability of the subtilase domain.[93] However, although these β domains are topologically similar to the bacterial subtilases, their sequences do not show any homology.[54,55,91] However, a clear sequence homology exists between the Tk-SP and KP-43 proteases, being 32% and 16% for the subtilisin and β jelly roll domains, respectively.[55] The β domains of the bacterial enzymes have binding sites for calcium ions as well as binding sites in their subtilase domains. In the case of Tk-SP, neither the β jelly roll domain nor is calcium binding to it required for activity or folding of the protein. It was shown, however, to be vital for the extreme thermal stability of the protease.[54,55] The calcium ions are tightly bound to the two binding sites in the β domain, as extensive 10 mM EDTA treatment is required to remove the ions from the molecule.[54,55] To determine the role of calcium binding for the thermal stability of Tk-SP, thermal denaturation of the intact form of the protein containing both subtilase and the β jelly roll domain was measured in the presence and absence of 10 mM $CaCl_2$ and compared with that of a truncated mutant of Tk-SP (Tk-S359ΔJ) where the β domain had been deleted and measured under the same conditions. In addition, the protein contained the mutation of active

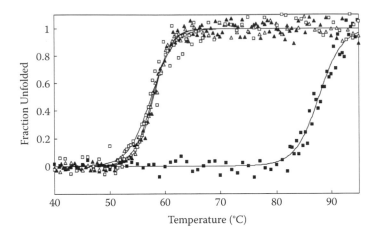

FIGURE 4.10   Thermal denaturation curves of Tk-S359A and Tk-S359AΔJ in the absence and presence of calcium. The thermal denaturation curves for Tk-S359A (filled squares) and Tk-S359AΔJ (filled triangles) were measured in 10 mM CaCl₂. For comparison are shown the curves for the same proteins, but in the absence of CaCl₂ and the presence of EDTA, Tk-S359A^{EDTA} (open squares), and Tk-S359AΔJ^{EDTA} (open triangles). The curves were recorded by monitoring the change in circular dichroism at 222 nm. (Reprinted from *Journal of Molecular Biology*, 400, T. Foophow et al., Crystal Structure of a Subtilisin Homologue, Tk-SP, from *Thermococcus kodakaraensis*, 865–877. Copyright 2010, with permission from Elsevier.)

site Ser to Ala.[55] The results of the thermal denaturation experiments were striking (Figure 4.10), showing that calcium binding to the jelly roll domain contributed more than 29°C to stabilization of the protein. Either deleting the domain (Tk-S359AΔJ and Tk-S359AΔJ^{EDTA}) or removing calcium ions from it (Tk-S359A^{EDTA}) decreased the T_m by 29.4°C to 29.5°C under the conditions used in those experiments.[55] These results show that the jelly roll domain, in the calcium-bound form, contributes significantly to the hyperthermostability of Tk-SP. Furthermore, on the basis of high sequence homology to subtilases from the hyperthermophilic archaea *P. furiosus*, *Thermococcus gammatolerans*, and *Thermococcus onnurineus*, it was suggested that swapping of the calcium-binding β jelly roll domain may be a mechanism of adaptation of hyperthermophilic proteases to their high-temperature environments.[55] There is also a significant sequence homology of the β domain of Tk-SP to that of the C-domain of AQUI from *T. aquaticus*.[54] Sequence alignments between these proteases suggest that one of the calcium sites (site 1) should also be present in the *Thermus* enzyme, although the 3D structure for that domain is not available for the enzyme. In the case of AQUI, however, there is no indication of a contribution by the putative β barrel C-domain to the thermostability of the enzyme,[94] its apparent biological function being that of promoting extracellular secretion of the protease.[39,40]

Aqualysin I, as with all other known subtilases from prokaryotes, contains specific calcium-binding sites that contribute significantly to the thermostability of these enzymes. However, considering the great structural similarity among the different subtilases, it is surprising to find the variability that exists among these enzymes in terms of their calcium-binding sites. The number of calcium-binding sites characterized so far within the subtilase domain of these enzymes range from none, as in Tk-SP, to seven, as in Tk-subtilisin. Apparently there are 16 different-calcium binding sites that have been identified in known crystal structures of subtilases. These occur both within and across homology families.

As discussed previously, the "true" subtilisins contain two calcium sites (sites A and B), Ca-1 and Ca-2. The Ca-1 calcium-binding site in subtilisin BPN′ is located in the loop at the C-terminus of helix C. It is formed as a residue 9 loop out of the last turn of the helix (Figure 4.8).[42,72] The calcium is coordinated by several residues in this loop comprising residues 75 to 83, in addition to a carboxylate group of Asp41 and the side of Gln2 at the N-terminus.[43,72,73,95,96] Interactions at this site tether the N-terminus of the molecule to the loops comprising residues 75 to 81 and 40 to 42.[96] The second binding site (site B or Ca-2) in the subtilisins is located in a shallow crevice between two loops consisting of residues 169 to 174 and 195 to 197.[95,96] This site is located 32 Å away from site 1 and appears to bind Ca^{2+} and other cations much more weakly.[97] The proposed role of these cation-binding sites in the folding of the subtilisins, and hence the contribution of calcium binding to the kinetic barrier of folding/unfolding, was discussed previously in this review. The role of calcium binding in thermodynamic and kinetic stabilization of the subtilisins has been well demonstrated in several studies.[43,72,85,86] Subtilisin binds calcium with a K_a of 10^7/M and 6.7×10^4/M to sites Ca-1 and Ca-2, respectively, at 25°C.[58] At a high concentration of calcium (100 mM), when calcium would occupy both binding sites, it was estimated that the binding energy contributes 13.5 kcal/mol to the free energy of unfolding.[72] Monovalent cations can also bind to the weaker binding site, albeit with a lower affinity, but contribute significantly to the stability of the native state in the absence of calcium ions. Stability of subtilisin BPN′ was shown to be enhanced by increasing the calcium-binding affinity of the weaker site by point mutations.[97] By introducing negatively charged groups, Pro172Asp and Gly131Asp, in the vicinity of the bound calcium, the binding affinity could be increased by about sixfold over that of the wild-type enzyme.[97]

Calcium sites 1 and 2 have also been identified in the crystal structures in related thermophilic subtilases thermitase from *T. vulgaris*[77,98] and Ak.1 protease (AkP) from *Bacillus* Ak.1.[79] Thermitase binds calcium ions with different affinities at three different sites in the enzyme.[77,98–100] Two of these sites correspond to the "strong" Ca-1 site and the third corresponds to the "weak" Ca-2 of the subtilisins. In addition, thermitase contains a third "medium-strength" binding site not present in the true subtilisins.[77,99,100] Binding to Ca-1 in thermitase is strong; even extensive incubation in the presence of 0.1 M EDTA did not remove it from the binding site.[100] The chelating ligands at this site are conserved to a considerable extent with respect to the subtilisins. There are some structural differences, however, that may contribute to stronger calcium binding at this site in the thermophilic enzyme. Thermitase has an N-terminal extension, not present in the subtilisins, that forms a number of contacts to the binding site. First, this extension provides the chelating side chain of Asp5, which replaces the carbonyl group of Gln2 with a carboxyl group, suggesting stronger binding as a result. Furthermore, the calcium-binding loop is a residue longer (Asn84) in thermitase as compared to subtilisins, which apparently provides a number of possible hydrogen bonds with the N-terminal extension.[77,99] The structure of the weaker calcium site in thermitase resembles that (Ca-2) of the subtilisins, but the medium-strength binding site was first described in the thermitase structure and is not present in the subtilisin BPN′ or Carlsberg. This calcium-binding site is located in a loop that leads into a helix (helix C in the subtilisins) containing the active site His at its N-terminus. The protein ligands are four carboxyl oxygens provided by residues Asp57, Asp60, and Asp62; the side-chain carbonyl of Gln66; and the main-chain carbonyl of Thr64.[77,99] The role of calcium binding to this site for the thermostability of thermitase has not been determined, but it is of interest that incorporation of this calcium-binding loop into the structure of subtilisin BPN′ by site-directed mutagenesis significantly stabilized the enzyme.[101] In the presence of 10 mM $CaCl_2$, the mutant enzyme containing the loop was 10 times more stable to irreversible thermal inactivation at 60°C than wild-type subtilisin BPN′. Insertion of the calcium loop affected the activity of the enzyme, as K_m for its standard substrate increased eightfold, although its k_{cat} was unaffected.[101] This change in K_m of the mutant may reflect the proximity of the loop to the active site and the substrate-binding site.

The thermostable AkP protease from *Bacillus* Ak.1 contains four calcium-binding sites, three of which correspond to the sites in thermitase.[79] The protease is markedly stabilized by calcium binding; for example, at 70°C, a 10-fold increase in Ca^{2+} concentration increased the half-life by three orders of magnitude.[102] The extent of stabilization by calcium binding is greater than for thermitase. In the absence of added calcium, the stability of AkP is slightly less stable than that of thermitase, with a half-life of less than 1 min at 80°C, compared with 9 min for thermitase. In the presence of Ca^{2+}, however, the half-life at 80°C was increased to 15 hr, compared with 19 min for thermitase.[79] This larger stabilizing effect of calcium on AkP protease is likely to be at least partly explained by the additional calcium-binding site in the enzyme. This site appears to be unique to AkP among the structurally characterized subtilases. It is situated close to the Ca-1 site. In fact, one residue, Glu83, is a ligand for Ca^{2+} at both sites; the carbonyl oxygen is ligand at site 1, whereas the carboxyl oxygens are ligands in the novel site.[79] Other ligands at this site are the carboxyl group of Asp50, the carbonyl oxygen of Pro47, and three water molecules. It has been proposed that this site acts cooperatively with the high-affinity Ca-1 site to give this enzyme additional stability compared with thermitase.[79,102]

Tk-subtilisin from the hyperthermophilic archaeon *T. kodakaraensis* contains seven calcium-binding sites.[48,103] The enzyme is highly thermostable, with a half-life of 9 hr at 90°C, and 50 min at 100°C.[104,105] The protein is highly stabilized by Ca^{2+} binding, but unlike other characterized bacterial subtilisins, it requires Ca^{2+} ions for folding, even in the presence of the Tk-propeptide.[48,49,103] In the absence of Ca^{2+} ions the protein folds into a molten globule-like structure. The high-affinity Ca-1 site of the bacterial subtilisins is conserved in the structure of Tk-subtilisin, but the other six sites are unique in the structure of the enzyme when compared to other known structures of subtilases.[48,49] The Ca-1 site is fully formed in the unautoprocessed pro-Tk-subtilisin, but is not formed late in the folding process, that is, during the autoprocessing, as must be the case for the bacterial subtilisins.[48,103] In the subtilisins, the site is not formed or is incompletely formed in the unautoprocessed form because one of the ligands, Gln2, can only coordinate with the Ca^{2+} ion after cleavage of the peptide bond (Tyr(-1)-Ala1) between the prodomain and the subtilase domain, and the new N-terminus structurally rearranges to place Gln2 in position to complete coordination of the Ca^{2+} ion at the Ca-1 site.[48] It is actually binding of Ca^{+2} to the site formed late in the folding process that is required to pull the equilibrium to the native conformation (Equation 4.1).[58] The fact that the Ca-1 and Ca-2 sites are formed so late in the folding process is believed to partly explain the large contribution of calcium binding to the unfolding energy barrier of the enzyme.[58,59] Comparison of crystal structures of the unautoprocessed form (pro-Tk-subtilisin) and that of the autoprocessed or mature forms of Tk-subtilisin indicated that the mature domain is almost completely folded before autoprocessing, including six of the calcium sites.[48] The seventh (Ca-7) calcium-binding site seems not to be fully formed in the unautoprocessed form, and a conformational change that is brought about by the autoprocessing reaction seems to favor the formation and stabilization of the Ca-7 site. Thus it was proposed that the Ca-7 site may have a similar role in Tk-subtilisin as described above for the Ca-1 site in the bacterial subtilisins; that is, by being formed or strengthened at the last stage of the folding reaction, binding of calcium to the site may help pull the folding reaction to the native state of the enzyme, and hence stabilize it.[48]

Four of the calcium-binding sites (Ca-2 through Ca-5) in Tk-subtilisin are located within a long inserted loop that is unique to this protein.[103] The loop is inserted between α helix 6 (α6m) and β strand 5 (β5m) in the protein (Figure 4.11), but these parts of the molecule, in addition to α helix 7 (α7m), form the αβα substructure that is believed to be crucial for the folding of bacterial subtilisins (Section 4.2.2.).[31,42,103] Incorporation of the calcium loop at this site might be required to promote the formation of this central αβα substructure as a folding nucleus in the folding of the entire protein at the extreme temperature conditions under which *T. kodakaraensis* grows.[103] Mutants of Tk-subtilisin, including a mutant, Δloop-Tk-subtilisin, in which the Ca^{2+}-binding loop was deleted, lost its ability to fold into a native structure.[103] Mutants in which the Ca-2 and Ca-3 sites were removed had slightly increased thermostability, so it was suggested that the Ca^{2+}-binding loop is required for folding of Tk-subtilisin, but does not seriously contribute to stability of the enzyme's native structure.[103] The Ca-6 site in Tk-subtilisin is also not found in other known structures of subtilases. It is located in a surface loop between helices 9 (α9m) and 10 (α10m) in the protein, and may contribute to tighter packing of these helices close to the C-terminus.[49] As the α10m helix interacts with the N-terminal part of the molecule, it was suggested that this site may be important for both thermal and proteolytic stability of Tk-subtilisin.[49]

WF146 protease, a subtilase from the thermophilic *Bacillus* sp. WF146,[106] shares a high degree of sequence identity with sphericase, a mesophilic subtilisin-like protease from *B. sphaericus*,[25,107] and two subtilisins (S39 and S41) from a psychrophilic Antarctic *Bacilli*.[108,109] Crystal structures of the meso- and psychrophilic enzymes show that both enzymes have five calcium-binding sites in their subtilase domain.[25,26] Furthermore, from sequence comparisons, the calcium-binding sites appear to be conserved in the thermophilic WF146 protease. None of the calcium-binding sites in these enzymes correspond to the high-affinity Ca-1 site of the subtilisins. Comparison

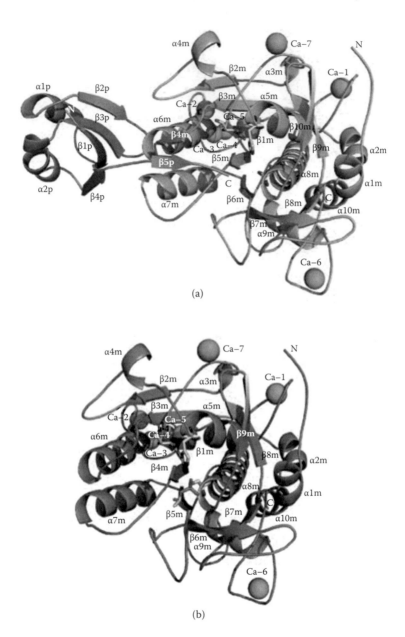

(a)

(b)

FIGURE 4.11 Three-dimensional structures of the (a) autoprocessed and (b) mature forms of Tk-subtilisin from *T. kodakaraensis*. For the structure of the autoprocessed form (a), the prodomain (p) and the mature (m) domain are colored in pink and green, respectively. The structure of the mature subtilase form (b) is colored orange. Active site residues are indicated by yellow sticks, and the seven calcium ions are shown as cyan spheres. (Reprinted from *Journal of Molecular Biology*, 372, S. Tanaka et al., Four New Crystal Structures of Tk-Subtilisins in Unautoprocessed, Autoprocessed, and Mature Forms, 1055–1069. Copyright 2007, with permission from Elsevier.) **(See color insert.)**

of the structures of the mesophilic sphericase (sph) and the psychrophilic subtilisin S41 revealed that the two structures are nearly identical and it was concluded that multiple calcium ion binding is not one of the molecular factors responsible for temperature adaptation of S41.[26] Since the crystal structure of the WF146 protease is not known, such structural comparison would not reveal the role of calcium loading to these sites for thermostabilization. It was shown, however, that deleting an insertion loop (residues 210 to 221) in WF146 protease, which is present in all of these enzymes and harbors calcium-binding site 2, significantly decreased the stability of the WF146 protease.[82] Furthermore, it was shown by the use of site-directed mutagenesis, that substituting residues in the vicinity of calcium-binding site 2 in WF146 protease to the corresponding residues in S41 significantly decreased the thermal stability of the thermophilic enzyme, which was attributed to weaker calcium binding at this site in the mutant.[82] Thus, in comparison with the cold-adapted enzyme, more favorable binding to calcium-binding site 2 in the thermophilic enzyme seems to contribute to its thermal stability.

Four calcium-binding sites have been identified in structures of subtilases belonging to the proteinase K family. Enzymes belonging to this family are highly dependent on calcium binding for thermal stability.[71,86,110,111] Proteinase K (PRK) has two calcium-binding sites, the stronger of the two corresponds to the Ca-2 site in the subtilisins, but the other, weaker binding site links loops at the N- and C-termini of the molecule by coordination of the calcium ion by residues Thr16 and Asp260, in addition to five water molecules.[24,71] The stronger binding site is present in both psychrophilic (*Vibrio* proteinase; VPR) and thermophilic (AQUI) representatives of the family.[21,38] The weaker calcium site is not present in these temperature extremophilic subtilases because of the absence of a carboxyl-containing residue at the C-terminus corresponding to Asp260 in PRK.[38,111] Both VPR and AQUI have a common calcium site close to the N-terminal part of the molecule, linking α helix A and residues of the succeeding loop (Figure 4.1). The Ca^{2+} ion is coordinated by the side chains and carbonyl oxygen of Asp9; the side chains of Asp12, Gln13, Asp19, and carbonyl oxygen of Asn21; and one water molecule.[21] This calcium site (termed Ca-3 in VPR) was first described in the structure of VPR, but it is present in several related enzymes of psychro-, meso-, and thermophilic origins.[21,80,112] The cold-adapted VPR also has a calcium-binding site, corresponding to the medium-strength calcium site of thermitase, described earlier. According to sequence alignments, this calcium-binding site is also present in highly homologous subtilases from other *Vibrio* species and a related enzyme from a psychrophilic *Serratia* sp., but is absent from AQUI and related thermophilic proteinases.[21] Thus, somewhat unexpectedly, in view of the role of calcium binding in stabilization of these proteinases, the psychrophilic enzyme contains an additional binding site when compared to the highly thermostable AQUI. These subtilases are highly dependent on Ca^{2+} binding for thermal stability (Figure 4.12). Under the conditions depicted in Figure 4.12, with or without 15 mM $CaCl_2$, but in the presence of 1 mM EDTA and 100 mM NaCl, $T_{50\%}$ (the temperature at which the enzyme loses half of its original activity in 30 min) in the absence of calcium decreased by ~26°C ($\Delta T_{50\%}$) for AQUI, ~12°C for PRK, and ~27°C for VPR (M. M. Kristjánsson, unpublished results). Furthermore, if NaCl was not present in the measuring buffer, $\Delta T_{50\%}$ increased even further, by ~40°C for the thermophilic

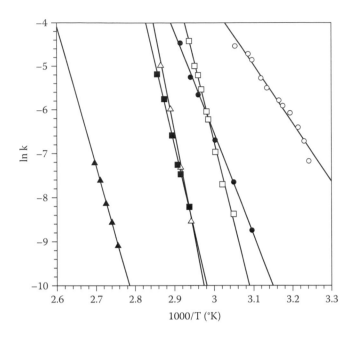

FIGURE 4.12   Effect of calcium binding on thermal inactivation of VPR (○), proteinase K (□), and AQUI (△). The enzymes were dissolved in 25 mM Tris-HCl, pH 8, containing 100 mM NaCl, 1 mM EDTA, and ±15 mM CaCl$_2$, were heated at selected temperatures, and the first-order rate constants of inactivation were determined at each temperature to construct the Arrhenius plots. Filled symbols represent measurements in the presence of calcium; open symbols represent measurements in the absence of calcium.

enzyme and ~23°C for PRK, while $T_{50\%}$ for VPR was not significantly affected. It is most likely that Na$^+$ ions bind to one or more of the calcium-binding sites of the proteins under the Ca^{2+}-free conditions. It has been shown that the thermostability of AQUI is dependent on calcium binding to at least two binding sites, with different affinities.[111] Of the two calcium-binding sites in AQUI, one of them corresponds to a strong and the other to a weaker binding site.[80] Binding of calcium to the weaker binding site appears crucial for thermostability of the enzyme, as removal of a Ca^{2+} ion from that site was detrimental for the stability of the enzyme.[111] It was suggested that the stronger calcium site corresponds to the Ca-1 site of PRK (corresponding to the weaker Ca-2 site in the subtilisins), but the weaker site was unidentified.[111] In the crystal structure of a PRK-like proteinase from a psychrotrophic *Serratia* sp. (SPRK), calcium occupied only the site corresponding to the novel Ca site (Ca-3) in VPR.[112] Structural alignments between VPR and SPRK strongly suggest that the other two calcium sites are also present in SPRK.[113] The fact that the Ca^{2+} ion occupies only this site, corresponding to the Ca-3 site in VPR, suggests that it may be the strongest calcium-binding sites in these subtilases, hence the stronger binding site in AQUI.[114] The crucial weaker binding site for thermostabilization of AQUI would therefore correspond to the Ca-1 site of PRK[24,71] or the Ca-2 site of the subtilisins.[95,96] It is of interest that this site in structures of different subtilases is frequently occupied by Na^+ or other monovalent ions if Ca^{2+} is not present. This has been shown, for example, for calcium-free PRK[71], AkP,[79] and the subtilisins.[72] AQUI is stabilized at high temperature in addition to Ca^{2+} by binding of different di- and trivalent metal ions to the weaker calcium-binding site.[111] The reason for the stabilizing effect of Na^+ on AQUI and PRK can most likely be similarly explained, that is, by binding to this same calcium-binding site in the proteinases.

Site-directed mutagenesis has been carried out on the Ca-3 site in VPR, with the objective of substituting the residues directly involved in calcium ligation to those in corresponding positions in AQUI. The sequences of the amino acid residues making up this calcium site in the two enzymes are identical except the three residues −Asp19Arg20Asn21− in VPR, of which residues 19 and 21 are ligands, are exchanged to −Ser19Asn20Ser21− in the thermophilic enzyme. Of these residues, the side chain of Asp19 and the carbonyl oxygen of Asn21 are ligands for Ca^{2+} at the site. A double mutant, Asp19Ser/Asn21Ser, and a triple mutant, Asp19Ser/Arg20Asn/Asn21Ser, of VPR were produced to simulate the binding site in AQUI, and their properties were determined with respect to activity and stability (M. M. Kristjánsson, unpublished results). Unexpectedly, both mutants had significantly diminished thermostability. For the double mutant, $T_{50\%}$ was decreased by ~8°C, but the Arg20Asn substitution appears to stabilize the mutant by ~5°C, hence the stability of the triple mutant decreased by ~3°C. Similar trends were observed for the stability of mutant enzymes that had been inhibited by phenylmethanesulfonyl fluoride (PMSF) and measured by thermal denaturation following changes in circular dichroism. The decrease in T_m for the double and triple mutant was ~6°C and ~4°C, respectively (M. M. Kristjánsson, unpublished results). These results underline the complex interplay of the calcium-binding sites and the other parts of the protein structure.

4.3.2.2   Other Structural Features

While it is clear that calcium binding is a major contributor to both kinetic and thermodynamic stabilization of the subtilases, other structural features intrinsic to the polypeptide chain are also at play in the thermostabilization of these enzymes. As for other thermozymes, the various mechanisms of thermostabilization, such as an increase in the number of hydrogen bonds, aromatic interactions, and salt bridges; stabilization of helix dipoles; improved packing of hydrophobic residues within core regions; shorter or tighter surface loops; and docking of N- and C-termini, all contribute to different extents to the stabilization of thermostable subtilases.[12,115–120] An understanding of the underlying molecular principles of the thermostabilization of proteins must be at the 3D structural level of the proteins.[120] Thus comparisons of homologous proteins of thermophilic organisms to those from meso- and psychrophiles can provide important insights into these mechanisms. On the basis of such comparisons of the crystal structures for the thermophilic subtilases, thermitase,[77,98] AkP protease,[79] AQUI,[80] Tk-subtilisin,[48] and Tk-SP[55] to those of structures of subtilases from psychrophilic[21,26,112] and mesophilic[24,53,70,81] organisms of the same homology families, one can attempt to establish the mechanisms of temperature adaptation of these enzymes. In addition, several crystal structures of thermostabilized mutants of the mesophilic subtilisins are available and can help to elucidate these mechanisms.[74]

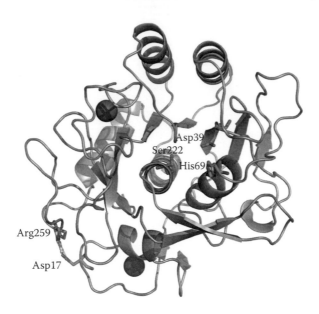

FIGURE 4.13 A ribbon diagram of the α-carbon backbone structure of AQUI, showing the catalytic triad residues and the putative salt bridge between residues Asp17 and Arg259. The bound calcium ions are shown as red spheres. **(See color insert.)**

Optimization of electrostatic interactions by increasing the number of salt bridges, often in networks or clusters, has been suggested to be a major contributor to thermostabilization, especially of hyperthermophilic proteins.[116,121–123] It may be misleading, however, when estimating the stabilization of proteins by salt bridges to simply compare their numbers between proteins. The contribution of salt bridges has been shown to be highly dependent on their structural context; the distance and orientation of the oppositely charged residues, their location in the protein and the electrostatic interaction between salt bridging side chains, with each other, as well as with its surroundings.[124,125] In fact, the desolvation penalty associated with bringing the separated charged groups of the unfolded protein from contact with water to form the ionic interaction in the folded protein may have a significant destabilizing contribution that must be paid off by the coulombic electrostatic interaction between the charged groups.[121,125]

Comparison of the crystal structures of VPR and AQUI suggested that the thermophilic enzyme may contain additional salt bridges in its structure that are not present in the psychrophilic counterpart. One such putative salt bridge is between residues Asp17 and Arg259 (Figure 4.13). The cold-adapted VPR contains Asn (Asn15) and Lys (Lys257) at the corresponding site in its structure.[126] Asn15 is located on the surface loop, which harbors the putative strong Ca-3-binding site in these enzymes, and points its side group toward the side chain of Lys257. By substituting Asn15 with Asp, a salt bridge could form between oppositely charged groups, which would mimic the putative salt bridge in the thermophilic proteinase. The Asn15Asp mutation did indeed increase the thermal stability of VPR by ~3°C, both in $T_{50\%}$ and T_m.[126] Recently we carried out the reverse mutation (i.e.,

Asp17Asn) on the thermophilic AQUI, thus deleting the putative salt bridge from its structure. Results of stability measurements showed that $T_{50\%}$ was decreased by 8°C to 9°C, indicating an important role for this salt bridge in the thermal stability of AQUI.[114]

Based on homology models of the 3D structures of two subtilases from hyperthermophilic archaea *P. furiosus* (pyrolysin) and *T. stetteri* (stetterolysin), and comparison to crystal structures of thermitase and subtilisin, it was suggested that higher thermostability of the hyperthermophilic subtilases could be correlated with an increased number of residues involved in ion pairs and such networks.[84]

Comparison of the crystal structures of thermitase and those of subtilisin BPN′ and savinase revealed that the thermophilic subtilase had a greater number of salt bridges than the mesophilic counterparts.[77,79] However, AkP protease did not have a greater number of salt bridges than the mesophilic subtilisin BPN′ and savinase, but the enzyme has four fewer salt bridges than thermitase, and it was suggested that it may be a factor in the higher intrinsic thermostability of the latter in the absence of calcium.[79] Both thermitase and AkP contain salt bridges that are expected to contribute to their thermostability. These salt bridges, which are located at the N- and C-termini of the proteins, act cooperatively with other local interactions to constrain the chain termini of the proteinases, so they are less prone to become sites of local unfolding.[79] In AkP, the α amino group of Trp1 forms a salt bridge with the side chain of Asp25, and its indole ring is oriented to pack against the pyrrolidine ring of Pro3, which helps to maintain it in an edge-to-edge interaction with the side chain of Trp24 in an aromatic interaction.[79] The C-terminal of the enzymes is similarly constrained by a double salt bridge between the carboxyl group and the side chain of Arg244, in addition to hydrogen bonding of the C-terminal Tyr to Gln251.[79]

In addition to the Trp1 and Trp24 aromatic interaction at the N-terminus, both thermitase and AkP contain a significantly greater number of aromatic interactions in their structures compared with their mesophilic counterparts.[77,79] In thermitase, 16 residues, representing two-thirds of the aromatic residues in the protein, were engaged in aromatic interactions. In contrast, only six residues, or one-third, were found to be involved in three such aromatic interaction pairs in subtilisin BPN′.[77] The additional aromatic clusters in these thermophilic subtilases are expected to contribute to their thermostability. Increased numbers of aromatic interactions have also been implicated in the stability of two hyperthermophilic subtilases, pyrolysin and stetterolysin, as judged from homology models of these enzymes.[84] However, their specific roles in stabilization have not been demonstrated.

The N-terminal extension of thermitase and AkP contain two prolines (Pro3 and Pro6). As prolines are limited in the number of configurations they can adopt, their presence in protein structures will restrict configurations at the site, and it is therefore expected that the presence of these prolines may help to constrain the configuration of the N-terminal region of proteins. Two proline residues (Pro5 and Pro7) near the N-terminus of the subtilase domain of AQUI may have a similar function in that enzyme. Structural comparisons of AQUI and the cold-adapted VPR showed that the thermostable enzyme has five proline residues that are not present in VPR, four of which are located in surface loops.[38] On the basis of this comparison between VPR and AQUI, prolines at these four sites in AQUI were substituted for those present at the corresponding sites in VPR and the properties of the mutants were measured.[127] The greatest effect on thermostability was observed for the Pro5Asn and Pro7Ile mutants; at 90°C, as the half-life for the Pro5Asn mutant was decreased from 45 min (in wild type) to 15 min in the mutant. For the Pro7Ile mutant, the effect on thermostability was even greater; the mutant was rapidly inactivated at 80°C, and even at 70°C, with half-lives of 10 min and 60 min, respectively. Melting temperatures measured by differential scanning calorimetry decreased from 94.0°C for wild type to 83.5°C and 75.7°C for the Pro5Asn and Pro7Ile mutants, respectively.[127] Reverse mutations have also been made on VPR, hence a single mutant, Ile5Pro, and a double mutant, Asn3Pro/Ile5Pro, were produced.[128] The mutations increased the thermostability of the enzyme by 5.7°C and 5.9°C in T_m and $T_{50\%}$, respectively, underscoring the importance of these Pro residues at the N-terminus in the structural stability of these enzymes.[127,128] An interesting effect of the proline substitutions on VPR was that the autoprocessing site at the N-terminus was shifted by two amino acid residues. In AQUI and enzymes containing Pro residues at these sites, the N-terminus is apparently constrained and anchored to the main body of the protein by a β sheet formed by the first two residues (Figure 14.4).[128] In the wild-type VPR, the N-terminus without the Pro residues is more flexible and fails to constrain it for interactions to the core enzyme. This apparently shifts the autoprocessing site by two residues, which in the more heat-stable enzymes participate in a β sheet, but lead to a more flexible N-terminal region in the cold-adapted subtilase.[128] The other extra proline residues present in AQUI, but not in VPR, are located on surface loops. An increased number of proline residues in surface loops are connected to thermostabilization of proteins.[117] An increased number of prolines are assumed to decrease the conformational entropy of the unfolded state more than for the folded state, thereby reducing the force of entropy-driven unfolding. The two single mutations of AQUI, Pro240Asn and Pro268Thr, decreased the T_m of the enzyme by ~2.2°C and 7.5°C, respectively.[127] The reverse single mutations, Asn238Pro and Thr265Pro, as well as the double mutant, were also found to be stabilizing when incorporated into the cold-adapted VPR.[129] Insertion of Pro residues into loops has been suggested as a stabilizing feature of the thermophilic WF146 protease when compared with its counterparts from meso- and psychrophiles.[82,106]

   Disulfide bonds stabilize proteins by insertion of cross-links into the structures, thereby decreasing the entropy of the unfolded state of the protein. The number of disulfides is quite variable between homology families of the subtilases, and these enzymes do not seem to rely highly on conserved disulfides for stabilization.[2] The mesophilic bacterial subtilisins do not contain disulfides, but a single disulfide bond is present in the psychrophilic subtilisins S39 and S41.[108,109] The presence of this disulfide bond does not seem to affect the stability of the enzymes.[108] A corresponding disulfide bond is also present in the related mesophilic enzymes, sphericase,[25] SSII,[107] the thermophilic WF146 protease,[82] and Tk-subtilisin.[48] The presence of this disulfide (Cys52-Cys65) was found to stabilize the WF146 protease against autolysis. The location of the disulfide bond within the structure gives support to this observation, as it is located on a surface loop that has previously been

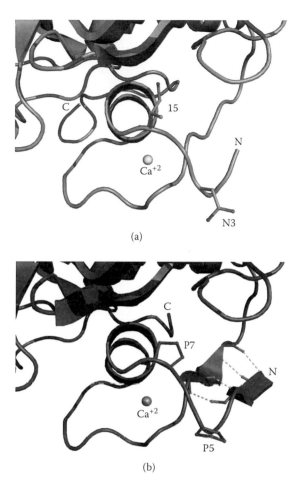

(a)

(b)

FIGURE 4.14　Comparison of the N-terminal regions of AQUI (red) and VPR (green). The mutated residues Asn5 and Ile5 and the proline residues (Pro5 and Pro7) at the corresponding sites in AQUI are shown as balls and sticks. (**See color insert.**)

shown to be the primary autolytic cleavage site in subtilisins.[101] Thus a disulfide cross-link at this location may provide these enzymes with resistance to thermally induced autolysis.

In addition to harboring a calcium-binding site, the Ca-2 loop in the structure of VPR is further enforced by a disulfide bond (Cys67-Cys99) that connects it to another loop carrying residues of the substrate-binding site of the enzyme.[21] This disulfide is also present in AQUI and related enzymes, and they also have a disulfide corresponding to Cys163-Cys194 in VPR in common.[21] Either, or both, of these disulfides are important for the structural integrity of the active enzymes, as the enzymes lost activity on reduction with dithiothreitol, even at 25°C. The loss of activity did not result from thermal denaturation of the proteinases, but merely by an increase in K_m (eightfold for VPR) of the enzyme against the substrate used for assaying the enzymes.[110] The role of these disulfides in AQUI have recently been studied by site-directed mutagenesis.[129] Single mutants, Cys99Ser and Cys194Ser, were produced, as well as a double mutant containing both mutations. Measurements of stability and the kinetic properties of the mutants indicated that Cys163-Cys184 is highly critical for both the stability and catalytic activity of the enzyme, more so than the Cys69-Cys99 bond. The T_m of the Cys194Ser mutant and the double mutant were 86.7°C and 86.9°C, respectively, compared with 94.0°C for the wild-type AQUI and 89.8°C for the Cys99Ser mutant.[129] It is noteworthy that the disulfides of these proteinase K-like subtilases are found in regions where other critical stabilizing features, such as calcium-binding sites and salt bridges, come together (see Figure 4.1).[21] It may be that the disulfides act as further enforcement at these sites to stabilize the native structure, or possibly in intermediates on the folding pathways of the enzymes. The AkP protease contains a disulfide bond located within its active site cleft. This disulfide appears to contribute to the integrity of the active site and increases the resistance of the enzyme against thermal inactivation.[79,102]

Several attempts have been made to engineer new disulfide bonds into the structure of subtilisin with the purpose of increasing their stability against thermal inactivation.[130–132] Many of these efforts did not result in stabilization of the enzymes. In one case, however, the equivalent of Cys61-Cys98 in AQUI was engineered into the structure of subtilisin E, with the effect that the half-life of the disulfide mutant was increased by two- to threefold and the T_m was increased by 4.5°C.[133] Another engineered disulfide that has been found to stabilize subtilisins, although not in their native form, is Cys3-Cys206, which was introduced into the structure of the Δ75-83 mutant by random mutagenesis. This disulfide cross-link would not form in the natural subtilisins, as the loop containing the Ca-1 site, which has been deleted in the Δ75-83 mutant, separates the N-terminus and the β strand containing residue 206. The inserted disulfide was found to slow down the half-life of thermal inactivation of the mutant by 17-fold.[13,74]

4.3.3 MUTAGENIC STUDIES

In 1988 Wells and Estell[133] wrote a short review entitled "Subtilisin—an enzyme designed to be engineered." Since then, the subtilisins have been extensively studied by both site-directed and random mutagenesis, with all of the amino acid positions having been modified by either of these approaches.[10] Besides the importance of subtilisins as industrial enzymes, other factors contributed to the research interest in these enzymes. The subtilisins were already important models for studies on enzyme mechanisms, and their timely cloning, ease of expression and purification, as well as availability of high-resolution crystal structures, all contributed to the important status of these proteins as a model enzyme system for protein engineering studies in the 1980s.[16] The subtilisins have been engineered by mutagenic methods with the purpose of modifying and gaining a better understanding of catalytic mechanisms, catalytic activities, substrate specificity, folding mechanisms, and protein stability under different conditions.[16]

Calcium-independent subtilisin variants have been designed that are more active and stable than the wild-type enzymes.[13,14] In designing these chelant-stable subtilisins, the Ca-1-binding loop was first deleted, as discussed previously (Figure 4.8), resulting in a mutant of low stability (Δ75-83), but which could fold without the IMC prodomain.[42-44] Through a combined approach of directed and random mutagenesis on the loop-deleted mutant and screening for increased stability, several compensating mutations were identified that restored stability when incorporated into the loop-deleted mutant. One such mutant, S88, containing 10 site-specific amino acid exchanges (Q2K, S3C, P5S, K43N, M50F, A73L, Q206C, Y217K, N218S, and Q271E), was stabilized (half-life at 65°C) 1000-fold under strongly chelating conditions (10 mM EDTA) and its activity was native-like.[44] The crystal structure of the S88 mutant has been determined and compared to the structure of subtilisin BPN′.[74] The comparison revealed that stabilization of the mutant could apparently be attributed to a large extent to a new disulfide bond formed between Cys3 and Cys206, as discussed previously, a new salt bridge of the amino group of Lys2 to the carboxyl of Asp41, in addition to two hydrogen bonds. Additional hydrogen bonds of Ser5 within the constrained N-terminus of the mutant are likely to explain the stabilizing effect of the P5S mutation. The Asp41 side chain is a ligand for calcium ion in the Ca-1 site of subtilisin BPN′, as the ionic interaction with Lys2 replaces the interaction to the calcium ion in the S88 mutant.[74] The mutations M50F, Y217K, N218S, and Q271E were previously known to be stabilizing[135] and were added to restore some of the lost stability of the loop-deleted mutant.[13,74] Strausberg et al.[14] were able to stabilize the S88 mutant further by site-directed random mutagenesis at 12 additional positions in the protein. Stabilizing mutations were identified at each of those sites, and collectively, using the optimal amino acid exchange at each of the 12 sites, the half-life of the protein was increased by 15,000-fold at temperatures of 65°C or greater. While there was not a direct correlation between stability and catalytic properties, some stabilized mutants also showed a higher catalytic efficiency over a range of selected substrates.[14]

Subtilases have also been the subject of several studies using directed evolution techniques.[15,17-20,136-139] Mesophilic subtilisin E has evolved into a thermostable enzyme functionally equivalent to that of the thermophilic homologue thermitase.[15] The mesophilic homologue SSII, or sphericase, has been cold adapted to a variant containing only four mutations, but which shows even more cold-adaptive traits than the psychrophilic subtilisin S41, with higher k_{cat} and lower thermal stability.[18] Subtilisin BPN′ has also evolved by such a directed evolutionary system into a cold-adapted mutant involving only three mutations.[137,138] Laboratory evolution methods have also been used to increase both the thermostability and catalytic activity of psychrophilic subtilisin S41.[17] By using three sequential rounds of random mutagenesis and recombinations, with screening of mutant libraries for increased thermostability without sacrificing the cold activity, it was possible to identify enzyme mutants with greatly enhanced thermostability and which were more active as well. A mutant containing mutations at seven positions had a half-life at 60°C about 500-fold that of the wild-type subtilisin S41, had its temperature optimum shifted upward by ~10°C, had its catalytic efficiency increased by a factor of three in the range 10°C to 60°C, and had its T_m shifted upward by 22°C.[17] By using five additional rounds of laboratory evolution, the stability of this mutant of S41 could be improved even further, such that the half-life at 60°C was increased by 1200-fold and the T_m was increased by 25°C compared with the wild-type enzyme.[19] In this mutant, which contained 13 amino acid substitutions, the catalytic efficiency and T_{opt} improved slightly. Of the

13 substitutions, only 2 are found within regular secondary structures, S78T and S252A, on α helices B and F (Figure 4.1); others are located in loop regions, based on the crystal structure of S41[26] (PDB entry 2gko). Two of those substitutions, N16D and Q69H, could result in additional salt bridges, the former possibly linking the N- and C-termini of the protease, as previously described for AQUI. It was note-worthy that 6 of the 13 mutations involved an exchange of a Ser to another amino acid.[19] There are suggestions from other studies that serines may be disfavored in thermophilic proteins, including subtilases.[19,38,84,129] This trend to exchange Ser for another amino acid in these laboratory-evolved enzymes, usually toward bulkier or more hydrophobic residues, may reflect selection observed in thermophilic evolution in natural proteins.[140]

Thermostability measurements in the presence of different calcium concentra-tions show that the stabilization of the mutants can largely be explained by enhanced calcium binding.[17,19] When the positions of the substitutions are examined within the now known crystal structure of S41, it is not unexpected that the calcium bind-ing of the enzyme was affected. Four of the mutations (K211P, R212A, D216E, and K221E) are present on a surface loop comprising residues 209 to 221, which harbors the Ca-2 calcium-binding site, and two other substitutions (N291I and S295T) are located on a loop consisting of residues 287 to 299, containing the Ca-1 site in the enzyme. Both of these calcium-binding sites were classified as high-affinity sites in the protein.[26] It should be noted that three of the apparently stabilizing mutations observed in the Ca-2 site in the laboratory-evolved mutants (Pro211, Ala212, Glu221) were found in the thermophilic homologue WF146 protease.[82] The reverse muta-tions (i.e., E221K and P211K/A212R) carried out on WF146 protease were found to significantly reduce the thermostability of the protease, and a mutant where this loop was deleted did not fold correctly.[82] Thus, instead of explaining the observed differences in the thermostability of these homologous proteases by the number of calcium-binding sites in their structures, reasons are to be found in the differences in amino acid composition of those sites, affecting the strength of calcium binding.[82] These results are a further example of the key role played by calcium binding in the thermostability of the subtilases.

REFERENCES

1. Rawlings, N. D., A. J. Barrett, and A. Bateman. 2010. MEROPS: The peptidase data-base. *Nucleic Acids Res* 38: D227–D233.
2. Siezen, R. J., and J. A. M. Leunissen. 1997. Subtilases: The superfamily of subtilisin-like serine proteases. *Protein Sci* 6: 501–23.
3. Siezen, R. J., B. Renckens, and J. Boekhorst. 2007. Evolution of prokaryotic subti-lases: Genome-wide analysis reveals novel subfamilies with different catalytic residues. *Proteins* 67: 681–94.
4. Fuller, R. S., A. Brake, and J. Thorner. 1989. Yeast prohormone processing enzyme (KEX2 gene product) is a Ca²⁺-dependent serine protease. *Proc Natl Acad Sci USA* 86: 1434–38.
5. Ottmann, C., R. Rose, F. Huttenlocher, A. Cedzich, P. Hauske, M. Kaiser, R. Huber, and A. Schaller. 2009. Structural basis for Ca²⁺-independence and activation by homodi-merization of tomato subtilase 3. *Proc Natl Acad Sci USA* 106: 17223–28.

6. Ekici, Ö. D., M. Paetzel, and R. E. Dalbey. 2008. Unconventional serine proteases: Variations on the catalytic Ser/His/Asp triad configuration. *Protein Sci* 17: 2023–37.

7. Bryant, M. K., C. L. Schardl, U. Hesse, and B. Scott. 2009. Evolution of a subtilisin-like protease gene family in the grass endophytic fungus *Epichloe festucae*. *BMC Evol Biol* 9: 168. doi:10.1186/1471-2148-9-168.

8. Tomkinson, B. 1999. Tripeptidyl peptidases: Enzymes that count. *Trends Biochem Sci* 24: 355–59.

9. Seidah, N. G., A. M. Khatib, and A. Prat. 2006. The proprotein convertases and their implication in sterol and lipid metabolism. *Biol Chem* 387: 871–77.

10. Maurer, K. H. 2004. Detergent proteases. *Curr Opin Biotechnol* 15: 330–34.

11. Saeki, K., K. Ozaki, T. Kobayashi, and S. Ito. 2007. Detergent alkaline proteases: Enzymatic properties, genes and crystal structures. *J Biosci Bioeng* 103: 501–8.

12. Kristjánsson, M. M., and B. Ásgeirsson. 2003. Properties of extremophilic enzymes and their importance in food science and technology. In *Handbook of Food Enzymology*, ed. J. R. Whitaker, A. G. J. Voragen, and D. W. S. Wong, 77–100. New York: Marcel Decker.

13. Strausberg, S. L., P. A. Alexander, D. T. Gallagher, G. L. Gilliland, B. L. Barnett, and P. N. Bryan. 1995. Directed evolution of a subtilisin with calcium-independent stability. *Biotechnology (NY)* 13: 669–73.

14. Strausberg, S. L., B. Ruan, K. E. Fisher, P. A. Alexander, and P. N. Bryan. 2005. Directed coevolution of stability and catalytic activity in calcium-free subtilisin. *Biochemistry* 44: 3272–79.

15. Zhao, H., and F. H. Arnold. 1999. Directed evolution converts subtilisin E into a functional equivalent of thermitase. *Protein Eng* 12: 47–53.

16. Bryan, P. N. 2000. Protein engineering of subtilisin. *Biochim Biophys Acta* 1543: 203–22.

17. Miyazaki, K., P. L. Wintrode, R. A. Grayling, D. N. Rubingh, and F. H. Arnold. 2000. Directed evolution study of temperature adaptation in a psychrophilic enzyme. *J Mol Biol* 297: 1015–26.

18. Wintrode, P. L., K. Miyazaki, and F. H. Arnold. 2000. Cold adaptation of a mesophilic subtilisin-like protease by laboratory evolution. *J Biol Chem* 275: 31635–40.

19. Wintrode, P. L., K. Miyazaki, and F. H. Arnold. 2001. Patterns of adaptation in a laboratory evolved thermophilic enzyme. *Biochim Biophys Acta* 1549: 1–8.

20. Arnold, F. H., P. L. Wintrode, K. Miyazaki, and A. Gershenson. 2001. How enzymes adapt: Lessons from directed evolution. *Trends Biochem Sci* 26: 100–106.

21. Arnórsdóttir, J., M. M. Kristjánsson, and R. Ficner. 2005. Crystal structure of a subtilisin-like serine proteinase from a psychrotrophic *Vibrio* species reveals structural aspects of cold adaptation. *FEBS J* 272: 832–45.

22. Ballinger, M. D., and J. A. Wells. 1998. Subtilisin. In *Handbook of Proteolytic Enzymes*, ed. A. J. Barrett, N. D. Rawlings, and J. F. Woessner, 289–94. London: Academic Press.

23. Kuhn, P., M. Knapp, S. M. Soltis, G. Ganshaw, M. Thoene, and R. Bott. 1998. The 0.78 Å structure of a serine protease: *Bacillus lentus* subtilisin. *Biochemistry* 37: 13446–52.

24. Betzel, C., S. Gourinath, P. Kumar, P. Kaur, M. Perbrandt, S. Eschenburg, and T. P. Singh. 2001. Structure of a serine protease proteinase K from *Tritirachium album* limber at 0.98 Å resolution. *Biochemistry* 40: 3080–88.

25. Almog, O., A. González, D. Klein, H. M. Greenblatt, S. Braun, and G. Shoham. 2003. The 0.93 Å crystal structure of sphericase: A calcium-loaded serine protease from *Bacillus sphaericus*. *J Mol Biol* 332: 1071–82.

26. Almog, O., A. González, N. Godin, M. de Leeuw, M. J. Mekel, D. Klein, S. Braun, G. Shoham, and R. L. Walter. 2009. The crystal structures of the psychrophilic subtilisin S41 and the mesophilic subtilisin Sph reveals the same calcium loaded state. *Proteins* 74: 489–96.

27. Ikemura, H., H. Takagi, and M. Inouye. 1987. Requirement of pro-sequence for the production of active subtilisin E in *Escherichia coli. J Biol Chem* 262: 7859–64.

28. Ikemura, H., and M. Inouye. 1988. *In vitro* processing of pro-subtilisin produced in *Escherichia coli. J Biol Chem* 263: 12959–63.

29. Eder, J., M. Rheinnecker, and A. R. Fersht. 1993. Folding of subtilisin BPN': Role of the pro-sequence. *J Mol Biol* 233: 293–304.

30. Eder, J., M. Rheinnecker, and A. R. Fersht. 1993. Folding of subtilisin BPN': Characterization of a folding intermediate. *Biochemistry* 32: 18–26.

31. Bryan, P., L. Wang, J. Hoskins, S. Ruvinov, S. Strausberg, P. Alexander, O. Almog, G. Gilliland, and T. Gallagher. 1995. Catalysis of a protein folding reaction: Mechanistic implication of the 2.0 Å structure of the subtilisin-prodomain complex. *Biochemistry* 34: 10310–18.

32. Zhu, X., Y. Ohta, F. Jordan, and M. Inouye. 1989. Prosequence of subtilisin can guide the folding of denatured subtilisin in an intermolecular process. *Nature* 339: 483–84.

33. Shinde, U. P., and M. Inouye. 1993. Intramolecular chaperones and protein folding. *Trends Biochem Sci* 18: 442–46.

34. Shinde, U., W. Fu, and M. Inouye. 1999. A pathway for conformational diversity in proteins mediated by intramolecular chaperones. *J Biol Chem* 274: 15615–21.

35. Marie-Claire, C., Y. Yabuta, K. Suefuji, H. Matsuzawa, and U. Shinde. 2001. Folding pathway mediated by an intramolecular chaperone: The structural and functional characterization of the aqualysin I propeptide. *J Mol Biol* 305: 151–65.

36. Chen, Y. J., and M. Inouye. 2008, The intramolecular chaperone-mediated protein folding. *Curr Opin Struct Biol* 18: 765–70.

37. Terada, I., S. T. Kwon, Y. Miyata, H. Matsuzawa, and T. Ohta. 1990. Unique precursor structure of an extracellular protease, aqualysin I, with NH$_2$ and COOH-terminal pro-sequences and its processing in *Escherichia coli. J Biol Chem* 265: 6576–81.

38. Arnórsdóttir, J., R. B. Smáradóttir, Ó. Th. Magnúsdóttir, S. H. Thorbjarnardóttir, G. Eggertsson, and M. M. Kristjánsson. 2002. Characterization of a cloned subtilisin-like serine proteinase from a psychrotrophic *Vibrio*-species. *Eur J Biochem* 269: 5536–46.

39. Kim, D. W., Y. C. Lee, and H. Matsuzawa. 1997. Role of COOH-terminal pro-sequence of aqualysin I (a heat-stable serine protease) in its extracellular secretion by *Thermus thermophilus. FEMS Microbiol Lett* 157: 39–45.

40. Kim, D. W., and H. Matsuzawa. 2000. Requirement for the COOH-terminal pro-sequence in translocation of aqualysin I across the cytoplasmic membrane in *Escherichia coli. Biochem Biophys Res Comm* 277: 216–20.

41. Takagi, H., M. Koga, S. Katsurada, Y. Yabuta, U. Shinde, M. Inouye, and S. Nakamori. 2001. Functional analysis of the propeptides of subtilisin E and aqualysin I as intramolecular chaperones. *FEBS Lett* 508: 210–14.

42. Bryan, P. N. 2002. Prodomains and protein folding catalysis. *Chem Rev* 102: 4805–15.

43. Bryan, P., P. Alexander, S. Strausberg, F. Schwarz, L. Wang, G. Gilliland, and D. T. Gallagher. 1992. Energetics of folding of subtilisin BPN'. *Biochemistry* 31: 4937–45.

44. Strausberg, S. L., P. A. Alexander, L. Wang, F. Schwarz, and P. Bryan. 1993. Catalysis of a protein folding reaction: Thermodynamic and kinetic analysis of subtilisin BPN' interactions with its propeptide fragment. *Biochemistry* 32: 8112–19.

45. Li, Y., Z. Hu, F. Jordan, and M. Inouye. 1995. Functional analysis of the propeptide of subtilisin E as an intramolecular chaperone for protein folding. *J Biol Chem* 270: 25127–32.

46. Ruvinov, S., L. Wang, B. Ruan, O. Almog, G. Gilliland, E. Eisenstein, and P. Bryan. 1997. Engineering the independent folding of the subtilisin BPN' prodomain: Analysis of two-state folding versus protein stability. *Biochemistry* 36: 10424–21.

47. Falzon, L., S. Patel, Y. J. Chen, and M. Inouye. 2007. Autonomic behavior of the propeptide in propeptide-mediated folding of prosubtilisin E. *J Mol Biol* 366: 494–503.
48. Tanaka, S., H. Matsumura, Y. Koga, K. Takano, and S. Kanaya. 2007. Four new crystal structures of Tk-subtilisins in unautoprocessed, autoprocessed, and mature forms: Insight into structural changes during maturation. *J Mol Biol* 372: 1055–69.
49. Tanaka, S., K. Saito, H. Chon, H. Matsumura, Y. Koga, K. Takano, and S. Kanaya. 2007. Crystal structure of unautoprocessed precursor of subtilisin from a hyperthermophilic archaeon. Evidence for Ca^{2+}-induced folding. *J Biol Chem* 282: 8246–55.
50. Tanaka, S., H. Matsumura, Y. Koga, K. Takano, and S. Kanaya. 2009. Identification of the interactions critical for propeptide-catalyzed folding of Tk-subtilisin. *J Mol Biol* 394: 306–19.
51. Gallagher, T., G. L. Gilliland, L. Wang, and P. Bryan. 1995. The prosegment-subtilisin BPN′ complex: Crystal structure of a specific "foldase." *Structure* 3: 907–14.
52. Ruan, B., J. Hoskins, L. Wang, and P. N. Bryan. 1998. Stabilizing the subtilisin BPN′ prodomain by phage display selection: How restrictive is the amino acid code for maximum stability? *Protein Sci* 7: 2345–53.
53. Jain, S., U. P. Shinde, Y. Li, M. Inouye, and H. Berman. 1998. The crystal structure of an autoprocessed Ser221Cys-subtilisin E-propeptide complex at 2.0 Å resolution. *J Mol Biol* 284: 137–44.
54. Foophow, T., S. Tanaka, Y. Koga, K. Takano, and S. Kanaya. 2010. Subtilisin-like serine protease from the hyperthermophilic archaeon *Thermococcus kodakaraensis* with N- and C-terminal propeptides. *Protein Eng Des Sel* 23: 347–55.
55. Foophow, T., S. Tanaka, C. Angakawidjaja, Y. Koga, K. Takano, and S. Kanaya. 2010. Crystal structure of a subtilisin homologue, Tk-SP, from *Thermococcus kodakaraensis*: Requirement of a C-terminal β-jelly roll domain for hyperthermostability. *J Mol Biol* 400: 865–77.
56. Li, Y., and M. Inouye. 1994. Autoprocessing of prothiosubtilisin E in which active-site serine 221 is altered to cysteine. *J Biol Chem* 269: 4169–74.
57. Fu, X., M. Inouye, and U. Shinde. 2000. Folding pathways mediated by an intramolecular chaperone. The inhibitory and chaperone functions of the subtilisin propeptide are not obligatory linked. *J Biol Chem* 275: 16871–78.
58. Fisher, K. E., B. Ruan, P. A. Alexander, L. Wang, and P. N. Bryan. 2007. Mechanism of kinetically-controlled folding reaction of subtilisin. *Biochemistry* 46:640–51.
59. Sari, N., B. Ruan, K. E. Fisher, P. A. Alexander, J. Orban, and P. N. Bryan. 2007. Hydrogen-deuterium exchange in free and prodomain-complexed subtilisin. *Biochemistry* 46: 652–58.
60. Cunningham, E. L., S. S. Jaswal, J. L. Sohl, and D. A. Agard. 1999. Kinetic stability as a mechanism for protease longevity. *Proc Natl Acad Sci USA* 96: 11008–14.
61. Salimi, N. L., B. Ho, and D. A. Agard. 2010. Unfolding simulations reveal the mechanism of extreme unfolding cooperativity in the kinetically stable α-lytic protease. *PloS Comp Biol* 6: 1–14.
62. Silen, J. L., and D. A. Agard. 1989. The α-lytic protease pro-region does not require a physical linkage to activate the protease domain *in vivo*. *Nature* 341: 462–64.
63. Sohl, J. L., S. S. Jaswal, and D. A. Agard. 1998. Unfolded conformations of α-lytic protease are more stable than its native state. *Nature* 395: 817–19.
64. Jaswal, S. S., J. L. Sohl, J. H. Davis, and D. A. Agard. 2002. Energetic landscape of α-lytic protease optimizes longevity through kinetic stability. *Nature* 415: 343–46.
65. Cunningham, E. L., and D. A. Agard. 2004. Disabling the folding catalyst is the last critical step in α-lytic protease folding. *Protein Sci* 13: 325–31.
66. Fuhrmann, C. N., B. A. Kelch, N. Ota, and D. A. Agard. 2004. The 0.83 Å resolution crystal structure of α-lytic protease reveals the detailed structure of the active site and identifies a source of conformational strain. *J Mol Biol* 338: 999–1013.

67. Truhlar, S. M. E., and D. A. Agard. 2005. The folding landscape of an α-lytic protease variant reveals the role of a conserved β-hairpin in the development of kinetic stability. *Proteins* 61: 105–14.

68. Truhlar, S. M. E., E. L. Cunningham, and D. A. Agard. 2004. The folding landscape of *Streptomyces griseus* protease B reveals the energetic costs and benefits associated with evolving kinetic stability. *Protein Sci* 13: 381–90.

69. Kelch, B. A., and D. A. Agard. 2007. Mesophile *versus* thermophiles: Insight into the structural mechanisms of kinetic stability. *J Mol Biol* 370: 784–95.

70. Takeuchi, Y., Y. Satow, K. T. Nakamura, and Y. Mitsui. 1991. Refined crystal structure of the complex of subtilisin BPN′ and *Streptomyces* subtilisin inhibitor at 1.8 Å resolution. *J Mol Biol* 221: 309–25.

71. Müller, A., W. Hinrichs, W. M. Wolf, and W. Sanger. 1994. Crystal structure of calcium-free proteinase K at 1.5 Å resolution. *J Biol Chem* 269: 23108–11.

72. Alexander, P. A., B. Ruan, and P. N. Bryan. 2001. Cation-dependent stability of subtilisin. *Biochemistry* 40: 10634–39.

73. Gallagher, T., P. Bryan, and G. L. Gilliland. 1993. Calcium-independent subtilisin by design. *Proteins* 16: 205–13.

74. Almog, O., D. T. Gallagher, J. E. Ladner, S. Strausberg, P. Alexander, P. Bryan, and G. L. Gilliland. 2002. Structural basis of thermostability. Analysis of stabilizing mutations in subtilisin BPN′. *J Biol Chem* 277: 27553–58.

75. Jakob, R. P., and F. X. Schmid. 2008. Energetic coupling between native-state prolyl isomerization and conformational protein folding. *J Mol Biol* 377: 150–75.

76. Jakob, R. P., G. Zoldák, T. Aumüller, and F. X. Schmid. 2009. Chaperone domains convert prolyl isomerases into generic catalysts of protein folding. *Proc Natl Acad Sci USA* 106: 20282–87.

77. Teplyakov, A. V., I. P. Kuranova, E. H. Harutyunyan, B. K. Vainshtein, C. Frommel, W. E. Hohne, and K. S. Wilson. 1990. Crystal structure of thermitase at 1.4 Å resolution. *J Mol Biol* 214: 261–79.

78. Gros, P., K. H. Kalk, and W. G. J. Hol. 1991. Calcium binding in thermitase. Crystallographic studies of thermitase at 0, 5 and 100 mM calcium. *J Biol Chem* 266: 2953–61.

79. Smith, C. A., H. S. Toogood, H. M. Baker, R. M. Daniel, and E. N. Baker. 1999. Calcium-mediated thermostability in subtilisin superfamily: The crystal structure of *Bacillus* Ak.1 protease at 1.8 Å resolution. *J Mol Biol* 294: 1027–40.

80. Barnett, B., P. R. Green, L. C. Strickland, J. D. Oliver, T. Ryder, and J. F. Sullivan. 2002. Aqualysin I, a serine protease from a member of the subtilisin superfamily: The crystal structure from an extreme thermophile, *Thermus aquaticus* YT-1. In *Program & Abstract Book*, vol. 29, p. 91. Buffalo, NY: American Crystallographic Association.

81. McPhalen, C. A., and M. N. G. James. 1988. Structural comparison of the serine proteinase-proteinase inhibitor complexes: Eglin-c-subtilisin Carlsberg and CI-2-subtilisin Novo. *Biochemistry* 27: 6582–98.

82. Bian, Y., X. Liang, N. Fang, X.-F. Tang, B. Tang, P. Shen, and Z. Peng. 2006. The roles of surface loop insertions and disulfide bond in the stabilization of thermophilic WF146 protease. *FEBS Lett* 580: 6007–14.

83. Yang, Y.-R., H. Zhu, N. Fang, X. Liang, C.-Q. Zhong, X.-F. Tang, P. Shen, and B. Tang. 2008. Cold-adapted maturation of thermophilic WF146 protease by mimicking the propeptide binding interactions of psychrophilic subtilisin S41. *FEBS Lett* 582: 2620–26.

84. Voorhorst, W. G. B., A. Warner, W. M. de Vos, and R. J. Siezen. 1997. Homology modelling of two subtilisin-like serine proteases from the hyperthermophilic archaea *Pyrococcus furiosus* and *Thermococcus stetteri*. *Protein Eng* 10: 905–14.

85. Voordouw, G., C. Milo, and R. S. Roche. 1976. Role of bound calcium ions in thermostable, proteolytic enzymes. Separation of intrinsic and calcium ion contributions to the kinetic thermal stability. *Biochemistry* 15: 3716–24.

86. Genov, N., B. Filippi, P. Dolashka, K. S. Wilson, and C. Betzel. 1995. Stability of subtilisins and related proteinases (subtilases). *Int J Peptide Protein Res* 45: 391–400.

87. Henrich, S., I. Lindberg, W. Bode, and M. E. Than. 2005. Proprotein convertase models based on the crystal structures of furin and kexin. *J Mol Biol* 345: 211–27.

88. Holyoak, T., M. A. Wilson, T. D. Fenn, C. A. Kettner, G. A. Petsko, R. S. Fuller, and D. Ringe. 2003. 2.4 Å resolution crystal structure of the prototypical hormone-processing protease Kex2 in complex with an Ala-Lys-Arg boronic acid inhibitor. *Biochemistry* 42: 6709–18.

89. Henrich, S., A. Cameron, G. P. Bourenkov, R. Kiefersauer, R. Huber, I. Lindberg, W. Bode, and M. E. Than. 2003. The structure of the proprotein processing proteinase furin explains its stringent specificity. *Nat Struct Biol* 10: 520–26.

90. Holyoak, T., C. A. Kettner, G. A. Petsko, R. S. Fuller, and D. Ringe. 2004. Structural basis for differences in substrate selectivity in Kex2 and furin protein convertases. *Biochemistry* 43: 2412–21.

91. Nonaka, T., M. Fujihashi, A. Kita, K. Saeki, S. Ito, K. Horikoshi, and K. Miki. 2004. The crystal structure of an oxidatively stable subtilisin-like alkaline serine protease, KP-43, with a C-terminal β-barrel domain. *J Biol Chem* 279: 47344–51.

92. Kobayashi, H., H. Utsunomiya, H. Yamanaka, Y. Sci, N. Katunuma, K. Okamoto, and H. Tsuge. 2009. Structural basis for the kexin-like serine protease from *Aeromonas sobria* as sepsis-causing factor. *J Biol Chem* 284: 27655–63.

93. Ueda, K., G. M. Lipkind, A. Zhou, X. Zhu, A. Kuznetsov, L. Philipson, P. Gardner, C. Zhang, and D. F. Steiner. 2003. Mutational analysis of predicted interactions between the catalytic and P domain of prohormone convertase 3 (PC3/PC1). *Proc Natl Acad Sci USA* 100: 5622–27.

94. Kurosaka, K., T. Ohta, and H. Matsuzawa. 1996. A 38 kDa precursor protein of aqualysin I (a thermophilic subtilisin-type protease) with a C-terminal extended sequence: Its purification and *in vitro* processing. *Mol Microbiol* 20: 385–89.

95. Bode, W., E. Papamokos, and D. Musil. 1987. The high-resolution x-ray crystal structure of the complex between subtilisin Carlsberg and eglin C, an elastase inhibitor from the leech *Hirudo medicinalis*. *Eur J Biochem* 166: 673–92.

96. Bott, R., M. Ultsch, A. Kossiakoff, T. Graycar, B. Katz, and S. Powers. 1988. The three dimensional structure of *Bacillus amyloliquefaciens* subtilisin at 1.8 Å and an analysis of the structural consequences of peroxide inactivation. *J Biol Chem* 263: 7895–906.

97. Pantoliano, M. W., M. Whitlow, J. F. Wood, M. L. Rollence, B. C. Finzel, G. L. Gilliland, T. L. Poulos, and P. N. Bryan. 1988. The engineering of binding affinity at metal ion binding sites for the stabilization of proteins: Subtilisin as a test case. *Biochemistry* 27: 8312–17.

98. Betzel, C., A. V. Teplyakov, E. H. Harutyunyan, W. Saenger, and K. S. Wilson. 1990. Thermitase and proteinase K: A comparison of the refined three-dimensional structures of the native enzymes. *Protein Eng* 3: 161–72.

99. Gros, P., Ch. Betzel, Z. Dauter, K. S. Wilson, and W. G. J. Hol. 1989. Molecular dynamics refinement of a thermitase-eglin-c complex at 1.98 Å resolution and comparison of two crystal forms that differ in calcium content. *J Mol Biol* 210: 347–67.

100. Frömmel, C., and C. Sander. 1989. Thermitase, a thermostable subtilisin: Comparison of predicted and experimental structures and the molecular cause of thermostability. *Proteins* 5: 22–37.

101. Braxton, S., and J. A. Wells. 1992. Incorporation of a stabilizing Ca^{2+}-loop into subtilisin BPN'. *Biochemistry* 31: 7796–801.

102. Toogood, H. S., C. M. Smith, E. N. Baker, and R. M. Daniel. 2000. Purification and characterization of Ak.1 protease, a thermostable subtilisin with a disulphide bond in substrate-binding cleft. *Biochem J* 350: 321–28.

103. Takeuchi, Y., S. Tanaka, H. Matsumura, Y. Koga, K. Takano, and S. Kanaya. 2009. Requirement of a unique Ca^{2+}-binding loop for folding of Tk-subtilisin from a hyper-thermophilic archaeon. *Biochemistry* 48: 10637–43.

104. Kannan, Y., Y. Koga, Y. Inoue, M. Haruki, M. Takagi, T. Imanaka, M. Morikawa, and S. Kanaya. 2001. Active subtilisin-like protease from a hyperthermophilic archaeon in a form with a putative prosequence. *Appl Environ Microbiol* 67: 2445–52.

105. Pulido, M., K. Saito, S. Tanaka, Y. Koga, M. Morikawa, K. Takano, and S. Kanaya. 2006. Ca^{2+}-dependent maturation of subtilisin from a hyperthermophilic archaeon, *Thermococcus kodakaraensis*: The propeptide is a potent inhibitor of the mature domain but is not required for its folding. *Appl Environ Microbiol* 72: 4154–62.

106. Wu, J., Y. Bian, B. Tang, X. Chen, P. Shen, and Z. Peng. 2004. Cloning and analysis of WF146 protease, a novel thermophilic subtilisin-like protease with four inserted surface loops. *FEMS Microbiol Lett* 230: 251–58.

107. Wati, M. R., T. Thanabalu, and A. G. Porter. 1997. Gene from tropical *Bacillus sphaericus* encoding a protease closely related to subtilisins from Antarctic bacilli. *Biochim Biophys Acta* 1352: 56–62.

108. Davail, S., G. Feller, E. Narinx, and C. Gerday. 1994. Cold adaptation of proteins. Purification, characterization, and sequence of the heat-labile subtilisin from the Antarctic psychrophile *Bacillus* TA41. *J Biol Chem* 269: 17448–53.

109. Narinx, E., E. Baise, and C. Gerday. 1997. Subtilisin from psychrophilic Antarctic bacteria: Characterization and site-directed mutagenesis of residues possibly involved in the adaptation to cold. *Protein Eng* 10: 1271–79.

110. Kristjánsson, M. M., Ó. Th. Magnússon, H. M. Gudmundsson, G. Á. Alfredsson, and H. Matsuzawa. 1999. Properties of a subtilisin-like proteinase from a psychrotrophic *Vibrio*-species. Comparison to proteinase K and aqualysin I. *Eur J Biochem* 260: 752–61.

111. Lin, S. J., E. Yoshimura, H. Sakai, T. Wakagi, and H. Matsuzawa. 1999. Weakly bound calcium ions involved in the thermostability of aqualysin I, a heat-stable subtilisin-type protease of *Thermus aquaticus* YT-1. *Biochim Biophys Acta* 1433: 132–38.

112. Helland, R., A. N. Larsen, A. O. Smalås, and N. P. Willassen. 2006. The 1.8 Å crystal structure of a proteinase K-like enzyme from a psychrotroph *Serratia* species. *FEBS J* 273: 61–71.

113. Larsen, A. N., E. Moe, R. Helland, and D. R. Gjellesvik. 2006. Characterization of a recombinantly expressed proteinase K-like enzyme from a psychrotrophic *Serratia* sp. *FEBS J* 273: 47–60.

114. Arnórsdóttir, J., M. Magnúsdóttir, Ó. H. Fridjónsson, and M. M. Kristjánsson. 2011. The effect of deleting a putative salt bridge on the properties of the thermostable subtilisin-like proteinase, aqualysin I. *Protein Peptide Lett* 18: 545–51.

115. Jaenicke, R., and G. Böhm. 1998. The stability of proteins in extreme environments. *Curr Opin Struct Biol* 8: 738–48.

116. Szilagyi, A., and P. Zavodsky. 2000. Structural differences between mesophilic, moderately thermophilic and extremely thermophilic protein subunits: Results of a comparative survey. *Structure* 8: 493–504.

117. Vielle, C., and G. J. Zeikus. 2001. Hyperthermophilic enzymes: Sources, uses, and molecular mechanisms for thermostability. *Microbiol Mol Biol Rev* 65: 1–43.

118. Chakravarty, S., and R. Varadarajan. 2002. Elucidation of factors responsible for enhanced thermal stability of proteins: A structural genomics based study. *Biochemistry* 41: 8152–61.

119. Razvi, A., and M. Sholtz. 2006. Lessons in stability from thermophilic proteins. *Protein Sci* 15: 1569–78.

120. Daniel, R. M., M. J. Danson, D. W. Hough, C. K. Lee, M. E. Peterson, and D. A. Cowan. 2008. Enzyme stability and activity at high temperatures. In *Protein Adaptation in Extremophiles*, ed. K. S. Siddiqui and T. Thomas, 1–34. New York: Nova Science.

121. Elcock, A. H. 1998. The stability of salt bridges at high temperatures: Implication for hyperthermophilic proteins. *J Mol Biol* 284: 489–502.

122. Xiao, L., and B. Honig. 1999. Electrostatic contributions to the stability of hyperthermophilic proteins. *J Mol Biol* 289: 1435–44.

123. Karshikoff, A., and R. Ladenstein. 2001. Ion pairs and the thermotolerance of proteins from hyperthermophiles: A "traffic rule" for hot roads. *Trends Biochem Sci* 26: 550–56.

124. Makhatadze, G. I., and P. L. Privalov. 1993. Contribution of hydration to protein folding thermodynamics. The enthalpy of hydration. *J Mol Biol* 232: 639–59.

125. Kumar, S., and R. Nussinov. 1999. Salt bridge stability in monomeric proteins. *J Mol Biol* 293: 1241–55.

126. Sigurdardóttir, A. G., J. Arnórsdóttir, S. H. Thorbjarnardóttir, G. Eggertsson, K. Suhre, and M. M. Kristjánsson. 2009. Characteristics of mutants designed to incorporate a new ion pair into the structure of a cold adapted subtilisin-like serine proteinase. *Biochim Biophys Acta* 1794: 512–18.

127. Sakaguchi, M., M. Matsuzaki, K. Niimiya, J. Seino, Y. Sugahara, and M. Kawakita. 2007. Role of proline residues in conferring thermostability on aqualysin I. *J Biochem* 141: 213–20.

128. Arnórsdóttir, J., Á.R. Sigtryggsdóttir, S. H. Thorbjarnardóttir, and M. M. Kristjánsson. 2009. Effect of proline substitutions on stability and kinetic properties of a cold adapted subtilase. *J Biochem* 145: 325–29.

129. Arnórsdóttir, J., S. Helgadóttir, S. H. Thorbjarnardóttir, G. Eggertsson, and M. M. Kristjánsson. 2007. Effect of selected Ser/Ala and Xaa/Pro mutations on the stability and catalytic properties of a cold adapted subtilisin-like serine proteinase. *Biochim Biophys Acta* 1774: 749–55.

130. Sakaguchi, M., M. Takezawa, R. Nakazawa, K. Nozawa, T. Kursakawa, T. Nagazawa, Y. Sugahara, and M. Kawakita. 2008. Role of disulfide bonds in a thermophilic serine protease aqualysin I from *Thermus aquaticus* YT-1. *J Biochem* 143: 625–32.

131. Wells, J. A., and D. B. Powers. 1986. *In vivo* formation and stability of engineered disulfide bonds in subtilisin. *J Biol Chem* 261: 6564–70.

132. Pantoliano, M. W., R. C. Ladner, P. N. Bryan, M. L. Rollence, J. F. Wood, and T. L. Poulos. 1987. Protein engineering of subtilisin BPN': Enhanced stabilization through the introduction of two cysteines to form a disulfide bond. *Biochemistry* 26: 2077–82.

133. Wells, J. A., and D. A. Estell. 1988. Subtilisin—An enzyme designed to be engineered. *Trends Biochem Sci* 13: 291–97.

134. Takagi, H., T. Takahashi, H. Momose, M. Inouye, Y. Maeda, H. Matsuzawa, and T. Ohta. 1990. Enhancement of the thermostability of subtilisin E by introduction of a disulfide engineered on the basis of structural comparison with a thermophilic serine protease. *J Biol Chem* 265: 6874–78.

135. Pantoliano, M. W., M. Whitlow, J. F. Wood, S. W. Dodd, K. D. Hardman, M. L. Rollence, and P. N. Bryan. 1989. Large increases in general stability for subtilisin BPN' through incremental changes in free energy of unfolding. *Biochemistry* 28: 7205–13.

136. Kuchner, O., and F. Arnold. 1997. Directed evolution of enzyme catalysts. *Trends Biotechnol* 15: 523–30.

137. Taguchi, S., A. Ozaki, and H. Momose. 1998. Engineering of a cold-adapted protease by sequential random mutagenesis and a screening system. *Appl Environ Microbiol* 64: 492–95.

138. Taguchi, S., A. Ozaki, T. Nonaka, Y. Mitsui, and H. Momose. 1999. A cold-adapted protease engineered by experimental evolution system. *J Biochem* 126: 689–93.

139. Pulido, M. A., Y. Koga, K. Takano, and S. Kanaya. 2007. Directed evolution of Tk-subtilisin from a hyperthermophilic archaeon: Identification of a single amino acid substitution responsible for low-temperature adaptation. *Protein Eng Des Sel* 20: 143–53.
140. Haney, P. J., J. H. Badger, G. L. Buldak, C. I. Reich, C. R. Woese, and G. J. Olsen. 1999. Thermal adaptation analyzed by comparison of protein sequences from mesophilic and extremely thermophilic *Methanococcus* species. *Proc Natl Acad Sci USA* 96: 3578–83.

5 Combined Computational and Experimental Approaches to Sequence-Based Design of Protein Thermal Stability

Julie C. Mitchell, Thomas J. Rutkoski,
Ryan M. Bannen, and George N. Phillips, Jr.

CONTENTS

5.1 INTRODUCTION

Protein design remains a challenge both for academic and industrial researchers looking to improve upon the products of natural evolution. Producing a protein with a particular function *de novo* is usually beyond our capabilities, and even redesigning an existing protein to enhance specific characteristics is challenging. One area of protein engineering in which success has been demonstrated is in the design of proteins exhibiting improved thermal stability. Proteins might be desired that have either longer functional lifetimes at some high operating temperature or improved activity at lower temperatures at the expense of thermal stability.

Thermostability has a fairly precise meaning in protein science and is different from the melting temperature or the temperature where the protein loses activity. Thermostability is defined as the difference in free energy between the folded and unfolded states, often satisfactorily defined with a two-state model. If the measurement of the fractional presence of an inactive conformation is a function of temperature, proteins often exhibit a sigmoidal melting curve, with the midpoint defined as the melting temperature. If the process is reversible, the fraction in the inactive (functionally unfolded) state can be related to the thermodynamic ΔG of unfolding by simple thermodynamic equations. In terms of a plot of ΔG as a function of temperature, shifting this curve to the right, broadening the curve, raising the curve, or any combination of these can increase the melting temperature. For proteins that do not show reversible, two-state melting curves where only the melting temperature can be determined, a kinetic measurement is used rather than a thermodynamic one.

Increases in the melting temperature of proteins, sometimes referred to as increased thermal stability (to differentiate from the more rigorous term thermostability) have been achieved by many methods. These include mimicking sequences from thermophilic organisms,[1] structure-based design,[2,3] selection screens,[4] or directed evolution.[5,6] All are valuable tools for the protein engineer.

Sometimes enzymatic or functional activity and thermal stability are intimately connected, and sometimes not. Activity is often lost concomitantly with a large melting transition. However, in the case of adenylate kinase from mesophilic and thermophilic *Bacillus* species, it was possible to maintain the broad mesophilic temperature activity range and increase the overall melting temperature by swapping the thermophilic protein's core into that of the mesophilic version.[7] The reverse was also possible, producing a protein that both unfolded easily and lost activity at lower than thermophilic temperatures.

A close connection between the range of secondary structures a given short polypeptide can adopt within a protein and the thermal stability of protein has been well established, and is referred to as the local structural entropy (LSE).[8] The average LSE over the entire protein sequence can be related to the stability of the protein and extended to design more stable proteins, as has been demonstrated with adenylate kinases.[9] This method has been further elaborated to allow more exhaustive use of sequence information in the design.[9] This method can be contrasted with the consensus method,[10] where one also aligns amino acid sequences of homologous proteins but simply takes the most commonly occurring amino acid at each position.

Currently there is great interest in improving the use of cellulose in the production of biofuels, particularly ethanol. Cellulose is a desirable feedstock for ethanol production, but current enzyme technology has not yet reached a point where the sugars can be broken down easily and economically on an industrial scale. One of the issues is that the long reaction times and high temperatures necessitate large amounts of cellulolytic enzymes (cellulases) in any industrial-scale process. Significant efforts by both government and industry have been directed toward improving the thermal stability of cellulases, though the problem has yet to be resolved for cellulosic ethanol production.

In the sections that follow, we describe both computational and experimental approaches to the design of thermally stable proteins. The computational methods attempt to optimize the LSE. In addition, we present ideas for how this optimization might be done without disrupting the secondary structure of the original protein. Some experiments designed to test the LSE method are described, as well as other experiments making use of directed evolution technology.

5.2 METHODS

5.2.1 IMPROVED CONFIGURATIONAL ENTROPY

The improved configurational entropy (ICE) model is a novel bioinformatic approach to design more thermally stable proteins. It does not require that the three-dimensional (3D) structure of the target protein be known, as in structure-based design, and it is less expensive and time-consuming than directed evolution. While the consensus method, a computational approach, also bypasses these issues, it requires that a large number of homologous sequences for the target molecule be known. The ICE method, in contrast, can be effectively utilized with as few as two protein sequences—a target and a homologue. This technique takes advantage of an empirical descriptor of LSE, which is based on structural information derived from the Protein Data Bank (PDB), as described in more detail below.

Local structural entropy is a value calculated from the observed structural diversity of an amino acid segment. This value was derived from careful analysis of the PDB[8] to examine how often certain amino acid tetramers appeared in protein secondary structures. A protein segment that can exist in multiple configurations is considered to have higher entropy than a sequence that is primarily found in a single configuration or secondary structure. Every possible configuration of a protein segment of four amino acids was analyzed and given an LSE value according to its observed diversity of secondary structures in the PDB.

There are correlations between thermal stabilities in a given family of proteins and the LSE scores. Based on this observation, the ICE method chooses amino acids from closely homologous proteins in order to replace existing sequences of the target sequence and reduce the overall structural entropy score. ICE proceeds by first aligning the homologous sequence(s) to a target protein sequence. The alignment reveals conserved residues, in which substitutions might affect folding or function, along with regions of variability in which the ICE program can perform substitutions. Tetramers that can be created at each position using amino acids from the aligned sequences are given LSE values as weights. An optimization technique (Dijkstra's method) is then used to assemble low-entropy tetramers into a complete sequence that has a minimal structural entropy score (Figure 5.1).

5.2.2 DIRECTED EVOLUTION AND SATURATION MUTAGENESIS

Directed evolution is a laboratory method based on the creation of a diversified library of related molecules through genetic manipulation and the subsequent identification of those variants within this population possessing optimized properties through selection or screening. Using the method of saturation mutagenesis described here, genetic sequences encoding a target protein are systematically mutated, then inserted

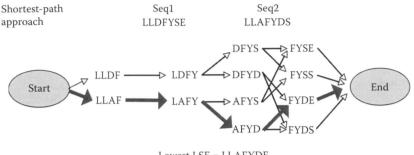

FIGURE 5.1 Example of the shortest-path implementation of ICE. **(See color insert.)**

into a bacterium for expression. The variants are then subjected to a screening procedure in which their functional activity is compared to the parental molecule. The sequence responsible for the greatest functional enhancement then serves as the new parental molecule and as a starting point for the subsequent round of diversification and screening. In performing multiple iterations of directed evolution, a protein can be selectively adapted for improved performance under conditions desirable for the researcher's intentions. This method has been previously explored with regard to improving enzyme stability.[11]

The gene encoding EngD (National Center for Biotechnology Information database P28623) was commercially synthesized after its sequence was codon optimized for expression in *Escherichia coli* by GeneArt (Regensburg, Germany). The gene was amplified by polymerase chain reaction (PCR) using Easy-A high-fidelity cloning enzyme (Stratagene, La Jolla, California) and subcloned into the pBAD/Thio TOPO vector (Invitrogen, Carlsbad, California). After transformation into TOP10 cells, the clone harboring the desired thioredoxin–EngD construct was identified by DNA sequencing. Expression and purification of EngD was completed as described previously, with several modifications.[12,13] The cells were lysed by sonication and the C-terminal His-tagged thioredoxin–EngD fusion protein was purified from the supernatant by immobilized nickel affinity chromatography. Fractions containing EngD, as determined by sodium dodecyl sulfate polyacrylamide gel electrophoresis (SDS-PAGE), were pooled and dialyzed overnight. Thioredoxin was cleaved from the fusion protein using EKMax recombinant enterokinase (0.5 U/mg of EngD; Invitrogen, Carlsbad, California). Following cleavage, the thioredoxin and free EngD were separated using a HiTrap Q HP anion exchange column (GE Healthcare, Piscataway, New Jersey). Pooled fractions exhibiting activity against the substrate 4-methylumbelliferyl β-D-cellobioside were dialyzed against 5 mM HEPES buffer, pH 7.0, containing NaCl (50 mM) and concentrated to 8.8 mg/ml. To confirm the enzymatic activity of EngD, purified protein preparations were assayed for their ability to hydrolyze the fluorogenic substrate 4-methylumbelliferyl β-D-cellobioside (MUC) (Sigma, St. Louis, Missouri).

Saturation mutagenesis was performed at select codons of the gene encoding for CelC based on the crystallographic B-factors of the corresponding amino acid residues (Figure 5.2). The program B-FITTER[14] was used to produce the rank listings

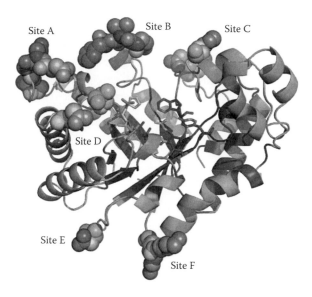

Site A   Site B   Site C

Site D

Site E

Site F

FIGURE 5.2 This model of *Clostridium thermocellum* CelC (1CEC) has been colored according to its B-factors (largest = red; smallest = blue). The atoms of the residues with the largest B-factors are shown as spheres. **(See color insert.)**

of residues with the highest B-factors from three crystallographic structures of CelC [PDB: 1CEN,[15] 1CEC,[16] and 1CEO.[15] Eight different sites were established that encompass those residues manifesting the 10 greatest B-factors in each of the above-mentioned crystallographic datasets. Site-directed mutagenesis was performed independently at each of these eight sites using 32-fold degenerate primer sets (NNK) or 16-fold degenerate primer sets (NDT/NDT) for those sites encompassing two codons. The resulting mutagenesis products were used to produce small libraries of *E. coli* expressing CelC in which one of the above sites has been randomized. Following growth, half of the culture was transferred to a separate glycerol stock plate for archival purposes.

Bacterial lysates were prepared from the remaining *E. coli* culture and transferred to a PCR plate. A 10 min thermal challenge was applied to the lysates that was sufficient to reduce the enzymatic activity of parental CelC by 70% to 90% (Figure 5.3). The enzymatic activity of the lysates was assayed by their ability to hydrolyze the fluorogenic substrate 4-methylumbelliferyl β-D-cellobioside. Residual enzymatic hits were identified as those wells whose residual activity exceeds a defined threshold (e.g., mean residual activity wild type + 2σ). Identified hits were retrieved from the master glycerol stock plates and rescreened using the same assay to confirm their status as hits. Plasmid DNA was isolated from confirmed hits and the *celC* gene was sequenced. Those amino acid substitutions responsible for the greatest increases in residual activity were "fixed" into the sequence and the iterative process was repeated at another site using this sequence as the new parental template for the subsequent round of saturation mutagenesis.

All oligonucleotides (including degenerate oligonucleotides) were obtained from Integrated DNA Technologies (Coralville, Iowa). Saturation mutagenesis, site-directed mutagenesis, and thermal challenges were performed using a BioRad DNA

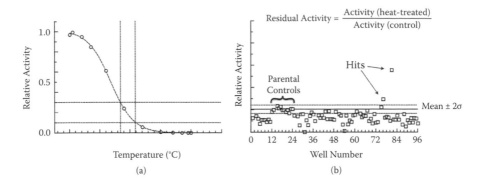

FIGURE 5.3 A thermocycler is used to apply a thermal challenge to crude cellular lysates that is sufficient to reduce parental CelC activity by 70% to 90% (a). Residual MUC activity is quantitated via a fluorescence-based assay (b). **(See color insert.)**

Engine (Hercules, California). Automated colony picking was performed using a Hudson Controls RapidPick Fully Automated High-Throughput Colony-Picking Workcell from Hudson Robotics Inc. (Springfield, New Jersey).

5.3   RESULTS

The enzyme adenylate kinase has been used in a number of studies to probe connections between structure, dynamics, and function. In terms of understanding or engineering the thermal stability of adenylate kinase based on primary sequence or tertiary structures, there have been several successful studies.[7,9] We will describe the application of a sequence-based approach (ICE) that has been successfully applied to adenylate kinase and is applicable to any protein where several closely related sequences are known. Additional studies for other proteins explore the use of a combinatorial approach based on high-throughput activity screens, as well as the application of a degenerate PCR method to explore the effects of single-site mutants on thermal stability of adenylate kinase variants. Finally, we present the application of ICE and protein structure analysis tools to engineer cellulase sequences that are predicted to enhance thermal stability while preserving protein secondary structure.

5.3.1   ADENYLATE KINASE

The first test of the ICE program used an exhaustive brute-force approach with the mesophilic adenylate kinase (AK) from *Bacillus subtilis* (AKmeso) as the target protein. Using only two sequences (a target and a homologue), we succeeded in making three variant proteins (AKlse1, AKlse2, and AKlse3) with higher thermal stability when compared with the wild-type AK.[9] Although this approach successfully demonstrated that local LSE could be used to fine-tune AK, the method was computationally expensive and was based on only two sequences, a target protein and a less stable homologue with very similar structures.[17] Applying ICE to sequences of three or more using this brute-force approach would be too computationally expensive.

Nonetheless, this provided evidence that the concept of inserting tetramers with lower local entropy from a homologue into a target sequence to increase the thermal stability of a protein was sound.

A second test was to be performed on adenylate kinase using three sequences: one target and two homologues. The brute-force approach required the creation of all possible sequences based on the allowable substitutions and the average LSE calculations for each substitution. The rate-limiting step was that each tetramer had to be looked up on a table 160,000 lines long. The most computationally expensive sequence, AKlse3, took 3 hr and 40 min to calculate all the possible entropy scores using this brute-force method with only two sequences. This method would be too computationally expensive when integrating three sequences into the procedure.

In order to solve this LSE minimization problem, it was modeled as a shortest-path network optimization problem.[9] This involved decomposing a protein sequence of length n into an ordered sequence of nodes in a network, each node representing a tetramer of amino acids. Because of the way the nodes are organized, there is no immediate upper limit on the number of allowable substitutions, and no practical limit for how many sequences can be incorporated in the algorithm. Dijkstra's algorithm,[18] which functions by solving one simple subproblem before expanding the subproblem, solving incrementally until the solution to the original problem is solved, was used to find the optimal sequence by finding the shortest path through this network from source to sink. In this case the shortest path was the lowest LSE value calculated through all the possible nodes of a sequence.

Using the shortest-path technique, the algorithm solved the problem for 19 amino acid positions and 3 sequences in less than 1 sec. This approach scales well to much larger problems. In a case of 10 sequences with 200 amino acids, each with 0% shared identity, the graph is on the order of 2 million nodes, but the global minimum LSE can be determined in about 24 hr.

A fourth AK variant (AKlse4) was created using the shortest-path algorithm.[19] Previously, AKmeso was the target sequence and AKpsychro was the homologue, but in this iteration AKthermo, an already highly thermally stable variant, was utilized as another homologue. It was hypothesized that AKlse4 would have higher stability than AKthermo because it would include not only the stability-enhancing mutations from the previous variants, but the mutations that give AKthermo its stability.

The shortest-path algorithm was successful in designing AKlse4 with the correct global LSE minimum in 0.8 sec. It was then expressed, purified, and analyzed for enzymatic activity. While the catalytic activity was maintained, its thermal stability was not improved over the other AKlse variants; instead, it was in the same range as AKlse1 and AKlse2 (Figure 5.4). The ICE program changed more than 20% of the target sequence, resulting in an enzyme that was not only more active, but more thermally stable compared to AKmeso, although not AKthermo.

5.3.2 CELLULASES

Improving the thermal stability of cellulases is a promising application of the ICE method. Ethanol production could be more cost effective with cellulases capable of withstanding higher temperatures, where enzymatic reactions are more efficient.

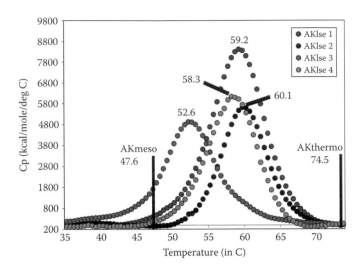

FIGURE 5.4 Differential scanning colorimetry (DCS) results for AKlse4 compared to the other variants and two wild-type AKs. **(See color insert.)**

While the computational algorithm of ICE was successfully expanded to include more sequences in the analysis, our first attempt at using the resulting sequences to produce a more stable cellulase was not successful. Working with the EngD/EngB cellulases from *Clostridium cellulovorans*, the first trial was performed using 40 mutations without using knowledge of the 3D structure of the EngD protein. This variant was expressed to some degree in soluble form, and the enzymatic activity was down by a factor of 30. After we completed the crystallization and structure determination (PDB ID:3NDY)[20] we mapped the mutations onto the structure, and it was clear that several of them were in regions of the protein likely to interrupt the binding of cellulose. Subsequently we scaled back the number of mutations by ruling out some that appeared to be in critical areas for activity and designed and expressed a protein with only 20 mutations. This sequence also expressed in soluble form, but not as well as the wild-type protein. The activity was also less than the wild type, but by a factor of about five, and it did not show improved thermal stability relative to the wild-type protein, indicating that the context of the mutations is also important.

Subsequently two thermally efficient cellulases were designed using a combination of ICE, Modeller,[21] Rosetta Design,[22] and PredictProtein.[23] In this experiment, cellulases were chosen from two thermophilic organisms, *Acidothermus cellulolyticus* and *Thermoascus aurantiacus*, as starting points. A protein BLAST search was performed and two homologous cellulases of 80% sequence identity were selected. Only the portion of the homologue that matched the alignment was used, and any gaps in the amino acid sequence were filled in with the amino acid of the original protein. These three sequences (the original and the two homologues) were then run through the ICE program to determine which tetramers would lower the overall LSE. Even though cellulases from thermophilic organisms were chosen, using the ICE program could still decrease the structural entropy of the starting cellulases, increasing their thermal stability.

A concern arose, however, because the ICE program does not constrain the likely secondary structure of the designed protein in relation to the original. For example, a β sheet may contain a tetramer that can be a part of many secondary structures (which would score high in terms of local entropy), and the ICE program might replace it with a tetramer that can only be found in α helices (a low local entropy score). This results in a net decrease in entropy (according to the ICE program), but it potentially changes the secondary structure, which could impact protein folding and stability. This might cause the ICE program to make changes that result in designing enzymes that are actually less stable, or even nonfunctional.

To check the extent to which the amino acid replacements distorted the secondary structures, Modeller was used to predict the 3D structures of the enzymes. After confirming that secondary structures were changed in several cases, and in one instance an α helix was added, steps were taken to reverse these structural changes.

First, Rosetta Design was used to design an enzyme using only the original enzyme and rotamers. This allowed, in essence, for a more accurate calculation of the energy of the original enzymes. This process was repeated for each of the original enzymes and also the enzymes designed by the ICE program. In cases where the secondary structures were deformed from the original, Rosetta Design returned significantly higher energy scores for the ICE-designed enzymes than the originals or their homologues. Next, Rosetta Design was used to revert the inappropriate choices that the ICE program made if the original amino acid provided a better folding energy. This procedure not only reversed questionable substitutions, but the final sequence had both lower LSE and Rosetta Design energies than the original cellulases from thermophilic organisms. The designed enzymes were thus predicted to be more stable based on a globally defined energy function and a locally defined measure of secondary structure propensity. In particular, it was found that although the Rosetta Design-modified enzymes did not score as well as the original ICE-designed enzymes, they have an improved LSE score in comparison with the original enzymes. Using PredictProtein and Modeller, the secondary structure and original protein were predicted to be similar. The secondary structure probabilities of the designed enzymes observed using PredictProtein were stronger than for the original enzyme, which is not surprising given that the ICE method selects for sequences with strong secondary structure preferences.

Ultimately this process allows one to balance the localized and "machine optimal" design achieved by ICE with more considerations related to the global fold in order to reduce structural entropy in a way that does not disrupt global protein architecture. Note that although ICE itself is a purely sequence-based method, the analysis using Rosetta Design and Modeller requires some structural information. In the absence of these structural details, a local reversal of unfavorable substitutions could be performed using information from PredictProtein, but the procedure might not reverse substitutions that disrupted tertiary contacts.

5.3.3 CₑₗC

In addition to computational studies on other systems, we explored a directed evolution approach to designing thermal stability in CelC. After an initial round of gene

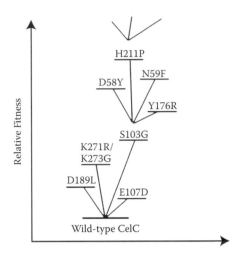

FIGURE 5.5 Iterative rounds of saturation mutagenesis were performed, and those amino acid substitutions that conveyed the greatest increase in fitness were "fixed" in the sequence and this new parental sequence was used for subsequent rounds of library generation and selection.

site saturation mutagenesis, four variants of CelC were found to exhibit a significantly higher residual activity following thermal challenge than did the wild-type enzyme: S103G, K271R/K273G, E107D, and D189L. Of these variants, the S103G substitution (site D in Figure 5.2) yielded the single greatest improvement in thermal tolerance. This variant was thus selected to be the new parental *celC* sequence used for the subsequent iteration of saturation mutagenesis.

Five variants were identified in the second round of mutagenesis (independent libraries were prepared corresponding to randomization of the amino acids at each of the sites in Figure 5.2): D58Y, N59F, Y176R, H211P, and H211W. The H211W substitution afforded the greatest enhancement in stability and so was again "fixed" in the sequence resulting in the S103G/H211W CelC variant that displayed a thermal tolerance several degrees higher than the wild-type enzyme after only two iterations of saturation mutagenesis. Subsequent rounds of saturation mutagenesis would be expected to yield further enhancement (Figure 5.5).

It should be noted that all the hits that are identified within a given round of iterative saturation mutagenesis cannot be assumed to result in an additive cumulative benefit when combined. This was found to be the case when the S103G/E107D/D189L/K271R/K273G CelC variant was prepared. Therefore it is essential to identify the substitution that results in the single greatest fitness improvement and then continue to select for additional gains from within the context of that new variant.

5.4 DISCUSSION

The success of the ICE method seems to depend on the degree of similarity of the sequences that are included in the LSE optimization. For adenylate kinase, with two proteins that were about 70% identical, the method worked well. However, for the EngD/B cellulase studies, the two proteins were not as homologous—only about

50% identical—and the LSE optimization suggested mutations that were not compatible with the 3D packing or secondary structure of the hybrid protein.

The results suggest that future work should consider whether attempting to switch a secondary structure type is likely to reduce the entropy and whether packing will be seriously disrupted by a proposed change. As described for two cellulases, postprocessing of the raw results from ICE can be performed to eliminate mutations that are expected to have adverse consequences based on these effects. In particular, we studied the combination of ICE and secondary structure prediction in order to design sequences that would improve thermal stability without disrupting the secondary structure of the original protein. In the future, it may be possible to directly incorporate this analysis into the ICE algorithm. The publicly distributed data on LSE simply provides the LSE score, which is based on the secondary structure preferences of a given tetramer in real proteins. By repeating the calculation of LSE scores and recording the dominant secondary structure type(s), it would be possible to build secondary structure preservation into the algorithm. In particular, in setting up the system of low-entropy tetramers, the method could be restricted to only those tetramers that frequently fold into the original protein's secondary structure. In addition to preserving the secondary structure, the algorithm would also be somewhat more efficient with this modification.

The iterative saturation mutagenesis strategy, which directly selects for the desired functionality of higher melting temperatures, seems more robust to the particular protein under study. However, it is much more expensive in terms of labor, materials, and required instrumentation. One can envision a combined approach in which computational models are used to limit the number of possibilities and experimental screening is used to test and evolve the designed sequences, which would be more effective than either approach alone.

ACKNOWLEDGMENTS

We would like to thank Syd Withers in the Microbial Synthetic Biology Center at the Great Lakes Bioenergy Research Center, University of Wisconsin–Madison, for his assistance in using the automated colony picker, as well as for helpful suggestions in designing the directed evolution screen. Plasmid pEZSeq-KanR containing *Clostridium thermocellum* endoglucanase *celC* was a kind gift from Dr. Phillip Brumm at C5·6 Technologies, Inc. (Middleton, Wisconsin). We also acknowledge the contributions of Robert W. Smith, Jeffrey Spence, and Euiyoung Bae to the research presented in this article. Finally, we thank Gerald Stoecklein for his work in helping to draft this manuscript. This work was supported financially by U.S. Department of Energy grants DE-FC02-07ER64494 and DE-FG02-04ER25627.

REFERENCES

1. Kristjansson, J. K. 1989. Thermophilic organisms as sources of thermostable enzymes. *Trends Biotechnol* 7(12): 349–53.
2. Bae, E., and G. N. Phillips, Jr. 2005. Identifying and engineering ion pairs in adenylate kinases. Insights from molecular dynamics simulations of thermophilic and mesophilic homologues. *J Biol Chem* 280(35): 30943–48.

3. Schweiker, K. L., and G. I. Makhatadze. 2009. A computational approach for the rational design of stable proteins and enzymes: Optimization of surface charge–charge interactions. *Methods Enzymol* 454: 175–211.

4. Counago, R., S. Chen, and Y. Shamoo. 2006. *In vivo* molecular evolution reveals biophysical origins of organismal fitness. *Mol Cell* 22(4): 441–49.

5. Giver, L., A. Gershenson, P.-O. Freskgard, and F. H. Arnold. 1998. Directed evolution of a thermostable esterase. *Proc Natl Acad Sci USA* 95(22): 12809–13.

6. Arnold, F. H. 1998. Design by directed evolution. *Acct Chem Res* 31(3): 125–31.

7. Bae, E., and G. N. Phillips, Jr. 2006. Roles of static and dynamic domains in stability and catalysis of adenylate kinase. *Proc Natl Acad Sci USA* 103(7): 2132–37.

8. Chan, C. H., H. K. Liang, N. W. Hsiao, M. T. Ko, P. C. Lyu, and J. K. Hwang. 2004. Relationship between local structural entropy and protein thermostability. *Proteins* 57(4): 684–91.

9. Bae, E., R. M. Bannen, and G. N. Phillips, Jr. 2008. Bioinformatic method for protein thermal stabilization by structural entropy optimization. *Proc Natl Acad Sci USA* 105(28): 9594–97.

10. Steipe, B. 2004. Consensus-based engineering of protein stability: From intrabodies to thermostable enzymes. *Methods Enzymol* 388: 176–86.

11. Eijsink, V. G., S. Gåseidnes, T. V. Borchert, and B. van den Burg. 2005. Directed evolution of enzyme stability. *Biomol Eng* 22(1–3): 21–30.

12. Hamamoto, T., O. Shoseyov, F. Foong, and R. H. Doi. 1990. A *Clostridium cellulovorans* gene, engd, codes for both endo-β-1,4-glucanase and cellobiosidase activities. *FEMS Microbiol Lett* 72(3): 285–88.

13. Murashima, K., A. Kosugi, and R. H. Doi. 2002. Thermostabilization of cellulosomal endoglucanase engB from *Clostridium cellulovorans* by *in vitro* DNA recombination with non-cellulosomal endoglucanase engD. *Mol Microbiol* 45(3): 617–26.

14. Reetz, M. T., and J. D. Carballeira. 2007. Iterative saturation mutagenesis (ISM) for rapid directed evolution of functional enzymes. *Nat Protoc* 2(4): 891–903.

15. Dominguez, R., H. Souchon, M. Lascombe, and P. M. Alzari. 1996. The crystal structure of a family 5 endoglucanase mutant in complexed and uncomplexed forms reveals an induced fit activation mechanism. *J Mol Biol* 257(5): 1042–51.

16. Dominguez, R., H. Souchon, S. Spinelli, Z. Dauter, K. S. Wilson, S. Chauvaux, P. Béguin, P. M. Alzari. 1995. A common protein fold and similar active site in two distinct families of beta-glycanases. *Nat Struct Biol* 2(7): 569–76.

17. Bae, E., and G. N. Phillips, Jr. 2004. Structures and analysis of highly homologous psychrophilic, mesophilic, and thermophilic adenylate kinases. *J Biol Chem* 279(27): 28202–8.

18. Dijkstra, E. W. 1959. A note on two problems in connexion with graphs. *Numer Math* 1: 269–71.

19. Bannen, R. M., V. Suresh, G. N. Phillips, Jr., S. J. Wright, and J. C. Mitchell. 2008. Optimal design of thermally stable proteins. *Bioinformatics* 24(20): 2339–43.

20. Bianchetti, C. M., R. W. Smith, C. A. Bingman, and G. N. Phillips, Jr. 2008. The structure of the catalytic and carbohydrate binding domain of endogluconase D. Protein Data Bank. Available at http://www.pdb.org/pdb/home/home.do. (PDB ID 3NDY.)

21. Eswar, N., B. Webb, M. A. Marti-Renom, M. S. Madhusudhan, D. Eraminan, M. Shen, U. Pieper, and A. Sali. 2006. Comparative protein structure modeling using Modeller. *Curr Protoc Bioinformatics* Unit 5.6.

22. Kuhlman, B., and D. Baker. 2000. Native protein sequences are close to optimal for their structures. *Proc Natl Acad Sci USA* 97(19): 10383–88.

23. Rost, B., G. Yachdav, and J. Liu. 2004. The PredictProtein server. *Nucleic Acids Res* 32 (Web Server issue): W321–26. Available at https://predictprotein.org.

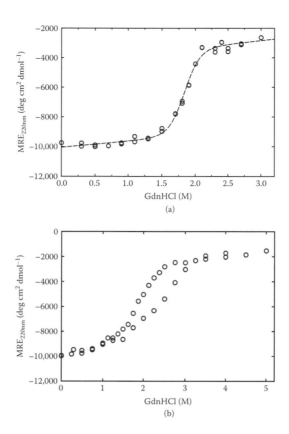

FIGURE 1.2 GdnHCl-induced equilibrium unfolding (blue circles) and refolding (red circles) curves of Tk-RNase HII at pH 9.0 for (a) 50°C for 2 weeks and (b) 20°C for 1 month. The reaction was followed by measuring CD at 220 nm. The broken line in (a) is the theoretical curve based on Equation (1.2). For the unfolding curve, Tk-RNase HII was incubated with different concentrations of GdnHCl. For the refolding curve, the protein, which was unfolded completely in 4 M GdnHCl, was diluted with buffer and the diluted protein solution was incubated.

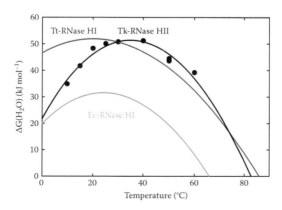

FIGURE 1.3 Stability curves for Tk-RNase HII (orange), Ec-RNase HI (cyan), and Tt-RNase HI (red) obtained from GdnHCl-induced equilibrium unfolding experiments. Filled circles are experimental data for Tk-RNase HII, and all lines represent the theoretical curves based on the Gibbs–Helmholtz equation (Equation 1.3).

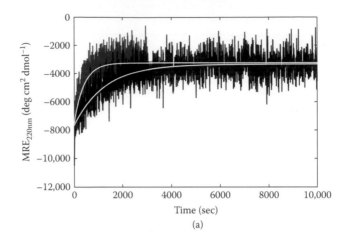

FIGURE 1.5 Kinetic experiments for Tk-RNase HII at 50°C and pH 9.0. (a) Kinetic unfolding curves for Tk-RNase HII monitored by a change in CD at 220 nm. The reaction was initiated by GdnHCl concentration jumps to 3.9 M (red line) and 3.4 M (black line).

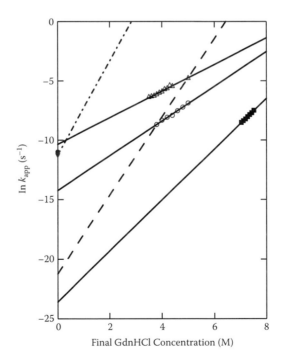

FIGURE 1.6 GdnHCl concentration dependence of $\ln k_{app}$ for unfolding of RNases H at 25°C. Open circles represent the data for Tm-RNase HII at pH 7.5; open triangles represent the data for Aa-RNase HII at pH 5.0; and filled squares represent the data for Sto-RNase HI at pH 3.0. The unbroken lines represent the linear fit. The long-dash line represents the data for Tk-RNase HII and the short-dash line represents the data for Ec-RNase HI.[25,37] The red circles and blue triangles represent the $k_{unf}(H_2O)$ value obtained from urea-induced unfolding experiments with Ec-RNase HI[38] and Tt-RNase HI,[39] respectively.

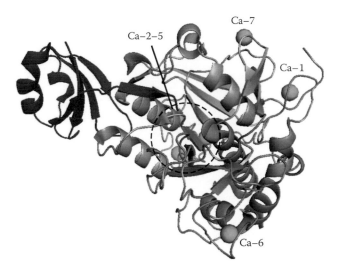

FIGURE 1.7 The crystal structure of mutant Tk-subtilisin mimicking its autoprocessed form. Propeptide and mature domain are represented in red and gray, respectively, and Ca²⁺ is shown in cyan.

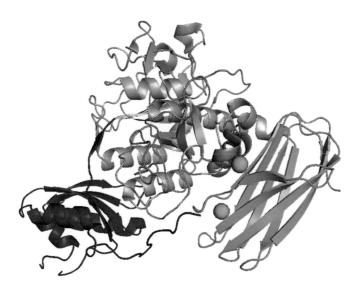

FIGURE 1.8 The crystal structure of the active site mutant of Pro-Tk-SP lacking a C-terminal propeptide. The N-terminal propeptide, mature domain, and extended C-terminal region are represented in red, gray, and green, respectively, and Ca²⁺ is shown in cyan.

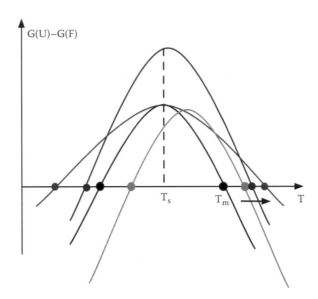

FIGURE 2.2 Stability curve of a given mesophilic protein (black curve) and possible modifications for the thermophilic homologue: up-shift (red curve), right-shift (orange curve), and broadening (magenta curve) of the stability curve. The two intercepts of each parabola with the axis correspond to equal populations of the folded and unfolded states and correspond to low (cold unfolding) and high (hot unfolding) temperatures. The intercepts are indicated with filled circles and the rightwise shift of the melting temperature, T_m, is indicated by the arrow.

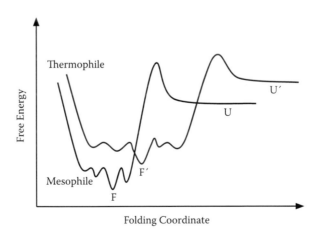

FIGURE 2.3 Pictorial representation of the free energy landscape of mesophilic and thermophilic homologues versus the folding coordinate. In this representation, the free energy basin of the thermophilic species is broader than its mesophilic counterpart.

FIGURE 2.4 Pictorial view of the electrostatic interaction in the protein matrix and distribution of the charged amino acids in the interior of the protein and at the interface with water. The charged residues are represented as red (negatively charged as Asp and Glu) and blue (positively charged as Lys and Arg) spheres. When buried in the interior of the protein, the charged amino acids tend to form an extended network of ion pairs or hydrogen bonds in such a way as to compensate for the desolvation free energy penalty. At the protein surface, the charged groups form either strong hydrogen bonds with water or the ion pair. According to Dardalat and Post,[46] the clustering at the interior of the protein gives rise to cohesive forces (arrows pointing to the interior of the core), while the distribution at the protein surface, via the coupling with interfacial water, favors adhesion to the solvent (arrows pointing toward the exterior).

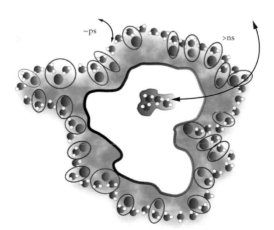

FIGURE 2.5 Organization of water around a protein. Water molecules form an extended network of hydrogen bonds surrounding the entire protein surface. The network is stabilized by local interactions with the hydrogen bond donor (blue spheres) and acceptor groups (red spheres) of the exposed amino acids. Depending on the composition of the protein surface, this layer is relatively stable with temperature. In the interior of the protein, polar or hydrophobic cavities may be filled with water molecules, and the water molecules are considered as structural elements of the protein matrix. While interfacial water molecules are quite mobile, with characteristic times slightly slower than bulk water molecules,[13,76] the internal water has reduced mobility and exchange with the bulk in nanoseconds or longer.[74,76]

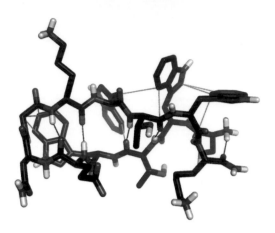

FIGURE 3.1 Constraint network of the tryptophan zipper 2 structure (PDB code 1le1). Covalent interactions are shown in black, hydrogen bonds are shown in magenta, and hydrophobic interactions are shown in green. Nitrogen atoms are colored in blue and oxygen atoms in red. Only polar hydrogen atoms are shown for clarity.

FIGURE 3.2 Rigid cluster decomposition of thermolysin-like protease (TLP), applied to constraint network representations of TLP at (a) 306, (b) 320, (c) 334, (d) 348, and (e) 362 K. Rigid clusters are shown as colored blobs, with the giant cluster colored in blue. Flexible regions are displayed as sticks.

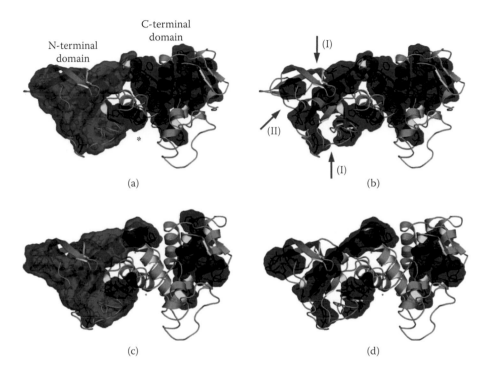

FIGURE 3.4 Snapshots from the thermal unfolding simulation of mesophilic TLP (a, b) and thermophilic thermolysin (c, d) just before (a, c) and after the phase transitions (b, d) at 350 and 373 K, respectively. The rigid cluster decomposition of the network is shown. The giant cluster is shown in blue. Other clusters are shaded in black. Arrows in (b) indicate potential unfolding nuclei. Roman numbers refer to the numbering of the unfolding nuclei. (*) The asterisk marks the active site. (Adapted from Radestock, S., and H. Gohlke, 2011, Protein Rigidity and Thermophilic Adaptation, *Proteins* 79: 1089–1108.)

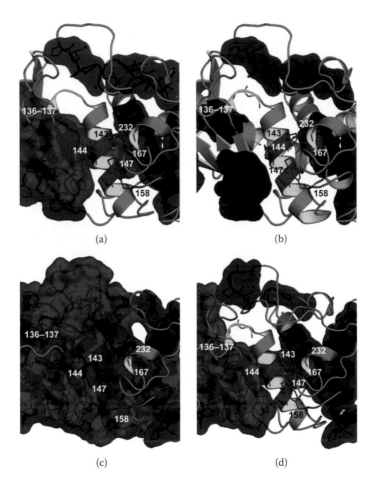

(a)　　　　　　　　　　　(b)

(c)　　　　　　　　　　　(d)

FIGURE 3.5　Active site of mesophilic TLP (a, b) and thermophilic thermolysin (c, d) at the working temperature of the mesophilic enzyme (342 K) (a, c) and the working temperature of the thermophilic enzyme (364 K) (b, d). The rigid cluster decomposition of the network is shown. The giant cluster is shown in blue. Other clusters are shaded in black. Catalytic residues are shown in red, while other functionally important residues are shown in green. (Adapted from Radestock, S., and H. Gohlke, 2011, Protein Rigidity and Thermophilic Adaptation, *Proteins* 79: 1089–1108.)

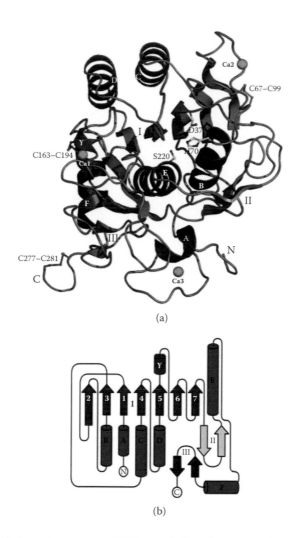

(a)

(b)

FIGURE 4.1 (a) Crystal structure of VPR, a subtilase from a psychrotrophic *Vibrio* sp. PA-44 (PDB entry 1sh7). The residues of the catalytic triad are shown in yellow, calcium ions are shown as green spheres, and disulfide bonds are in orange. (b) A topology diagram of the VPR structure. (From J. Arnórsdóttir, M. M. Kristjánsson, and R. Ficner, Crystal Structure of a Subtilisin-Like Serine Proteinase from a Psychrotrophic *Vibrio* Species Reveals Structural Aspects of Cold Adaptation, 2005, *FEBS Journal* 272: 832–845. Copyright Wiley-VCH Verlag GmbH & Co. KGaA. Reproduced with permission.)

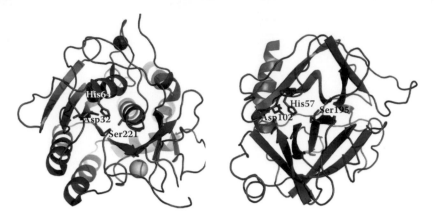

FIGURE 4.2 Schematic representations of typical 3D structures of serine proteases belonging to the subtilase (gray) and chymotrypsin-like (blue) superfamilies. The structures are those of subtilisin Carlsberg (PBD entry 1scn) and bovine chymotrypsin (PDB entry 2gmt). Labeled are residues of the catalytic triad, following the order DHS for subtilases and HDS for the chymotrypsin-like enzyme. Bound calcium ion to subtilisin is shown as a red sphere and natrium ion is in yellow.

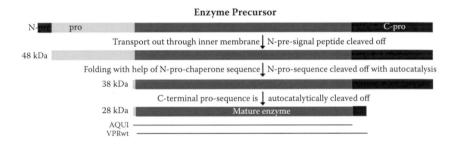

FIGURE 4.3 Processing of precursor proteins of VPR, a psychrotrophic subtilase from *Vibrio* sp. PA-44, and AQUI from the thermophile *Thermus aquaticus* YT-1 upon folding and secretion. A signal peptide is shown in red, the N-terminal prodomain is in yellow, the C-terminal domain is in green, and the mature subtilase domain is in gray. The mature AQUI is 2 residues longer at the N-terminus and 15 residues shorter at the C-terminus than the wild-type VPR (VPRwt).

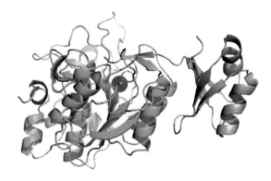

FIGURE 4.4 The structure of prodomain:subtilisin BPN complex (PDB entry 1spb). The subtilisin domain is shown in green and the residue 77 prodomain is shown in brown.

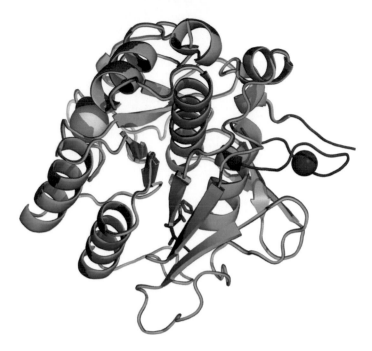

FIGURE 4.8 Superposition of ribbon diagrams of the α-carbon backbone of subtilisin BPN (PDB entry 1sud) (red) and the Δ75-83 mutant (PDB entry 1suc) (green). Calcium ions are shown as red spheres and a potassium ion occupying the Ca-2 site in the Δ75-83 mutant is shown in yellow. Residues of the active site are shown in pink.

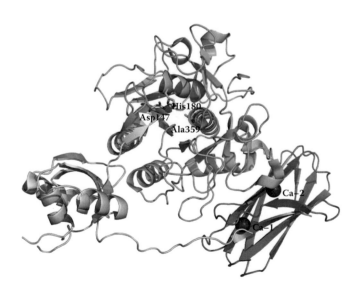

FIGURE 4.9 Backbone structure of ProN-Tk-SP (ProN-TK-S359A) from *T. kodakaraensis* (PDB entry 3afg). The N-prodomain (Lys4-Ala113), subtilase domain (Val114-Tyr421), and C-terminal β jelly roll domain (Ala442-Pro552) are shown in yellow, green, and blue, respectively. The catalytic triad residues Asp147, His180, and Ala359 (active site Ser359Ala mutant) are indicated and the two calcium ions (Ca-1 and Ca-2) of the C-domain are shown as red spheres.

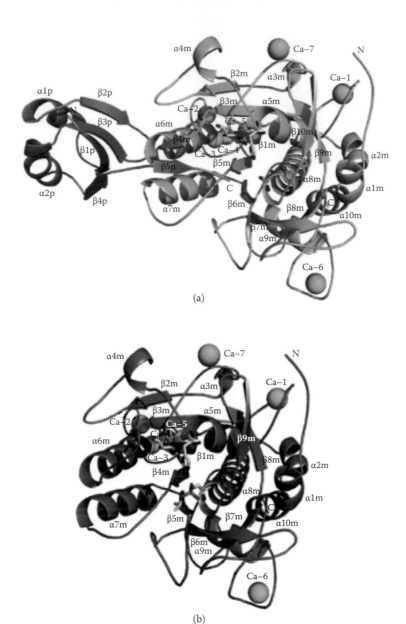

(a)

(b)

FIGURE 4.11 Three-dimensional structures of the (a) autoprocessed and (b) mature forms of Tk-subtilisin from *T. kodakaraensis*. For the structure of the autoprocessed form (a), the prodomain (p) and the mature (m) domain are colored in pink and green, respectively. The structure of the mature subtilase form (b) is colored orange. Active site residues are indicated by yellow sticks, and the seven calcium ions are shown as cyan spheres. (Reprinted from *Journal of Molecular Biology*, 372, S. Tanaka et al., Four New Crystal Structures of Tk-Subtilisins in Unautoprocessed, Autoprocessed, and Mature Forms, 1055–1069. Copyright 2007, with permission from Elsevier.)

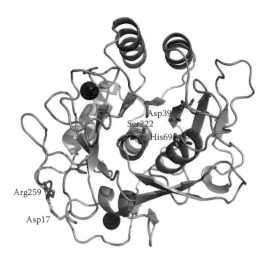

FIGURE 4.13 A ribbon diagram of the α-carbon backbone structure of AQUI, showing the catalytic triad residues and the putative salt bridge between residues Asp17 and Arg259. The bound calcium ions are shown as red spheres.

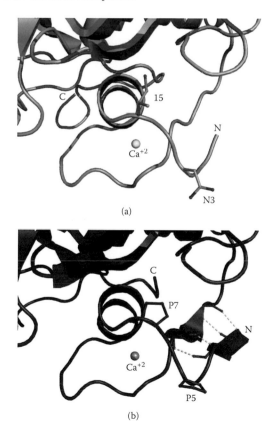

FIGURE 4.14 Comparison of the N-terminal regions of AQUI (red) and VPR (green). The mutated residues Asn5 and Ile5 and the proline residues (Pro5 and Pro7) at the corresponding sites in AQUI are shown as balls and sticks.

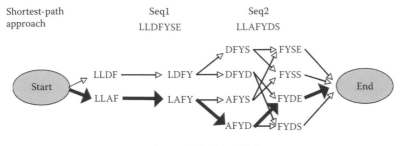

Lowest LSE = LLAFYDE

FIGURE 5.1 Example of the shortest-path implementation of ICE.

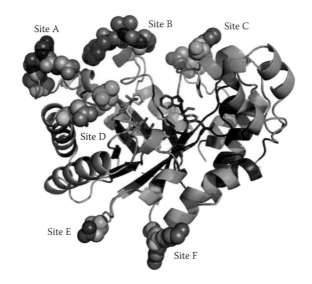

FIGURE 5.2 This model of *Clostridium thermocellum* CelC (1CEC) has been colored according to its B-factors (largest = red; smallest = blue). The atoms of the residues with the largest B-factors are shown as spheres.

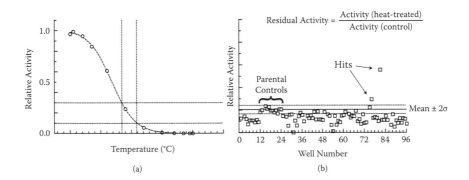

FIGURE 5.3 A thermocycler is used to apply a thermal challenge to crude cellular lysates that is sufficient to reduce parental CelC activity by 70% to 90% (a). Residual MUC activity is quantitated via a fluorescence-based assay (b).

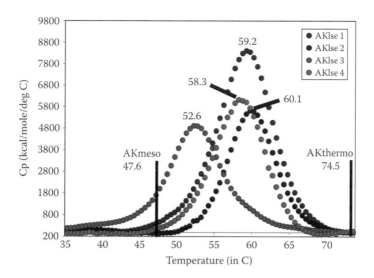

FIGURE 5.4 Differential scanning colorimetry (DCS) results for AKlse4 compared to the other variants and two wild-type AKs.

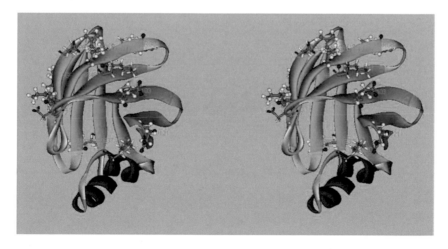

FIGURE 6.4 Stereo view of the 3D structure of the designed protein for the case of the crystal structure 2ifb with the key residues and mutated residues highlighted as sticks and balls and sticks, respectively.

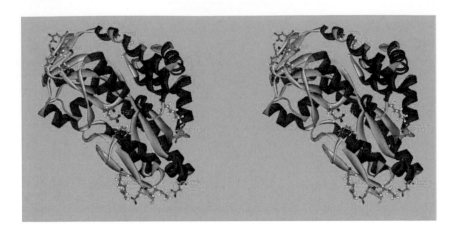

FIGURE 6.7   Stereo view of the 3D structure of the designed protein for the case of the crystal structure 2v4c with the key residues and mutated residues highlighted as sticks and balls and sticks, respectively.

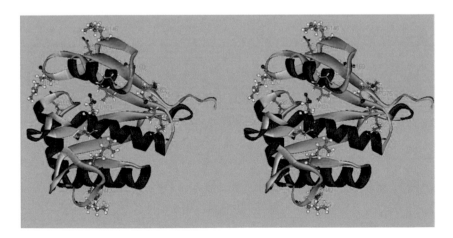

FIGURE 6.10   Stereo view of the 3D structure of the designed protein for the case of the crystal structure 2bl1 with the key residues and mutated residues highlighted as sticks and balls and sticks, respectively.

6 Designing Thermophilic Proteins

A Structure-Based Computational Approach

Sohini Basu and Srikanta Sen

CONTENTS

6.1 INTRODUCTION

Many natural microorganisms that are inhabitants of extreme environments like volcanic systems, highly acidic environments, and alkaline lakes, among others, have been discovered over the last decade. Such organisms are commonly known as extremophiles.[1] Extremophiles are generally classified on the basis of the extreme conditions in which they live. Some of the extremophiles known so far are thermophilic organisms (living at high temperatures), psychrophilic organisms (inhabitants of very low temperatures), acidophiles (living at extreme acidic pH), alcalophiles (living at extreme basic pH), and halophilic organisms (living in the presence of high salt concentrations). The discovery of such extremophiles is of great importance not only for their potential use, but also because these organisms and their biomolecules are valuable sources for understanding the physical basis of their extraordinary properties. In fact, proteins and enzymes isolated from extremophiles are considered useful for a variety of applications due to their extraordinary properties to thrive in conditions traditionally regarded as hostile. In this chapter we restrict our discussion to thermophilic proteins.

Thermophilic organisms are found in hot places such as hot springs and hot vents at the bottom of the sea. Depending on the optimal growth temperature (T_{opt}) thermophilic organisms are divided into two classes: moderate thermophiles ($50°C < T_{opt} < 80°C$) and hyperthermophiles ($T_{opt} > 80°C$). Given the fact that thermophilic proteins are made of the same building blocks as mesophilic proteins, it is important to understand the molecular basis of the mechanisms through which thermophilic proteins achieve their enhanced stability and maintain their activities at temperatures of 100°C or more. Extensive research has been conducted on the physicochemical origin of thermostability, and these studies have suggested that a variety of factors can contribute to this observed enhanced stability.[2–12] Such factors include optimized electrostatic interactions, improved packing, networks of hydrogen bonds, hydrophobic interactions, and cavity filling.[4,6,7,13–20] The enhancement of thermostability in thermophilic proteins is now believed to be the consequence of a number of locally improved interactions in three dimensions. For more detailed accounts of the physical basis of thermostability, see Chapters 1 through 4 in this book.

The possible important biotechnological applications of proteins that are capable of functioning at high temperatures have attracted much interest in understanding the mechanisms of thermophilicity and designing tailor-made thermophilic proteins. Biotechnological applications of normal enzymes are often limited due to their low heat stability. Enzymes isolated from thermophilic species are natural examples of proteins that are stable at high temperatures,[21,22] and they have the potential for use in numerous industrial applications. The most well-known example of a successful application of a thermophilic enzyme is Taq DNA polymerase isolated from *Thermus aquaticus*.[23,24] This enzyme has allowed the automation of polymerase chain reaction (PCR) technology that is widely used in research laboratories and industries all over the world. Amylase is another example of a thermophilic enzyme that has widespread biotechnological applications for the production of glucose and xylanase to whiten paper.[1]

Thus, both from a fundamental point of view as well as from the perspective of industrial applications, it is highly desirable to develop methods for designing thermostable variants of selected normal mesophilic proteins without altering their three-dimensional (3D) structures and functions significantly. Successful design of thermophilic protein is not only a technological and industrial requirement, it is also an indicator of the level of our understanding of the mechanisms of thermostability.

Several design approaches have already been developed for making proteins more thermally stable.[25–30] In general, these methods depend heavily on sequence comparisons. One of the approaches is directed evolution, which attempts to accelerate natural evolution in a laboratory setting.[31] A number of random mutations are first made to the original protein sequence. These resulting mutants are evaluated for improvement of a predefined specific trait. The mutants that perform the best are then used for additional cycles of mutations until the mutant variant with the desired property is obtained. There are success stories reporting designed proteins with improved thermostability using this approach.[5,27,31,32] However, this approach has its own limitations: for example, directed evolution requires a significant amount of laboratory resources and thus is expensive and time-consuming.

Another approach is the consensus method, which is a computational approach to improve thermal stability. In this method, a multiple sequence alignment is first performed using the target sequence and a large number of homologous sequences in order to find commonality among proteins within the same family.[26,33,34] If a majority of the homologous sequences have the same amino acid in a particular position and if it is found to be different from the amino acid in the target sequence, the consensus method states that the amino acid in the target sequence should be mutated to the amino acid shared by the majority of the homologous sequences. In this method, when there is not a clear majority amino acid bias at a given position in the sequence alignment, it becomes difficult to pick up the amino acids to mutate. Moreover, the consensus method requires that a large number of homologous sequences be known for the target protein.

More recently, a method has been developed that uses sequence alignments to redesign proteins to be more stable through optimization of local structural entropy.[35,36] Using this method, more stable variants of a mesophilic adenylate kinase with only the sequence information of one psychrophilic homologue were designed. Interestingly, the redesigned proteins display considerably improved thermal stabilities while still retaining catalytic activity.[37] Dantas et al.[29] reported that using their computational method Rosetta Design, they successfully designed nine different proteins that showed a dramatic increase in their stabilities in experiments. Rosetta Design has two main components, an energy function that ranks the relative fitness of various amino sequences for a given protein structure, and a search function for rapidly scanning sequence space. Rosetta Design uses a simple Monte Carlo optimization to identify low-energy sequences. Starting from a completely random sequence, single amino acid substitutions are accepted or rejected using the Metropolis criterion.

There is also a structure-based design approach that involves making particular amino acid mutations to specifically improve traits in the protein's structure. These mutations can be made to improve van der Waals interactions, hydrogen bonds, salt bridges, and interactions with ions and disulfide bridges, among others.[30] Structure-based design has been shown to successfully increase the thermal stability of proteins in a number of cases.[5,27,30,38] This method is limited by the fact that the 3D structure of the protein must be known before the mutations can be chosen. However, despite recent advances in computational protein design,[39,40] we do not have a complete understanding of the basic principles that govern the design and selection of naturally occurring proteins.[41] While several attempts to design proteins with a desired fold have been successful,[40,42] rational design of proteins with desired thermal properties is still an elusive goal, mostly because of the inability to predict the correct 3D structure starting from only the amino acid sequence. In that sense, use of experimentally known 3D protein structures in designing tailor-made proteins seems to be an easy way out.

In the present work we have utilized current knowledge on the mechanisms of enhanced thermostability to develop a computational method for designing a thermophilic protein starting with the 3D structure of a mesophilic protein through several key mutations. In order to remain consistent with the current beliefs that electrostatic interactions are the most important factors in enhancing the stability of a thermophilic protein, we increase the number of charged/polar residues on the surface of the selected protein by mutating several preselected surface residues. There are two key aspects of the current approach: (1) to identify the polar or charged residues that are associated with low energies of interaction with the rest of the protein and are close to the surface of the protein, and (2) for each such residue, finding another residue in its vicinity and replacing it with a matching conformer of a suitable residue with an electrostatic nature complementary to the key residue selected in the previous step in order to enhance its interaction with neighboring residues of the protein without any interresidue steric conflicts and thus stabilize the protein further. In this chapter, we present three case studies to demonstrate the capabilities of the present method. The enhanced stability of the designed protein in one case was further tested by performing molecular dynamics (MD) simulations at high temperatures. We have also validated the methods by comparing similar features of four known natural thermophilic–mesophilic protein pairs, including a very well-studied pair. The results indicate that the present approach is an easy and efficient computational method that can be successfully employed to convert any protein into a thermophilic one through several rationale-based mutations. Our designing method differs from the other methods in several unique ways: (1) we do not use sequence comparison for finding the desired mutations; (2) the protein's 3D features and interresidue interactions are explicitly considered at the atomic level; (3) desirable mutations are identified on the basis of interaction energies to stabilize tertiary contacts; (4) we predict mutations by the specific nature of residues at the single-residue level using realistic side-chain conformers; and (5) it allows designing of a thermophilic protein simply by mutating a minimal number of preselected residues. In principle, the present approach should be applicable to any protein with any kind of complex 3D architecture, even though in this chapter we have considered only single-chain single-globule proteins for simplicity. Finally, as experimental verification of the designed mutant proteins is beyond our scope, we have used computational methods to demonstrate the enhanced stability of the engineered (mutated) protein compared with the original mesophilic one. Preliminary results of such an approach have already been published.[43]

6.2 METHODS

6.2.1 Basic Rationales of Our Approach

In our design approach, we used our knowledge of mechanisms of enhanced stability in thermophilic proteins. Recent works have indicated that hyperthermophilic proteins are characterized by an increased number of surface charges and ionic networks that play dominant roles in enhancing thermostability.[8,44,45] It was further demonstrated that the desolvation cost for the formation of a salt bridge decreases at higher temperatures, leading to effective stabilization.[46] Thus it appears that ion-pair interactions should play a major role in enhancing the overall stability of thermophiles. The importance of the dynamic arrangement of ion pairs and their individual contributions to the overall thermal stability of a protein have recently been demonstrated by MD simulation of the crystal structures of thermophilic and mesophilic proteins.[47] Moreover, in proteins there are two types of H-bond-forming residues, those that are charge neutral and those with a charge, and thus there may be H-bonds between charged–neutral and neutral–neutral residues. It has been suggested that the desolvation energy for making an H-bond is lower than that for an ion pair, and the binding energy of a charged–neutral H-bond is far larger than that from neutral–neutral H-bonds due to the charge–dipole interaction.[48] These imply that charged residues might have an additional influence on stability through the formation of H-bonds. All these together indicate a strong influence of electrostatic interaction in enhancing the overall stability of a thermophilic protein. On the other hand, hydrophobic interactions prevent unfavorable aqueous solvation and also improve van der Waals interactions by cavity filling. Hydrophobic interaction is another factor that is believed to stabilize proteins at physiological temperatures. However, the importance of hydrophobic interaction in enhancing the stability of protein structures at higher temperatures is still debatable.[49] Hence, in the present design strategy, we have chosen modulation of the electrostatic interactions and the van der Waals interactions in improving the overall thermostability of a protein.

We further consider it quite likely that the thermal denaturation of a protein starts at the protein surface. Openings are created at the protein surface due to thermal fluctuation. Solvent molecules enter into the protein body, inducing more denaturation of the inside of the protein as the process continues. Therefore, making the surface of a protein more resistant to large thermal fluctuations should be a good strategy to prevent denaturation of the protein at a given temperature. Based on this concept, our basic strategy is to mutate a set of preselected less stable surface residues to achieve enhanced stability. There is literature[50–54] demonstrating that residues at the surface play important roles in enhancing the thermostability of known thermophilic proteins. Recently, the crystallographic group at Riken (Wako, Japan) found that the CutA1 protein isolated from a hyperthermophile has an extremely high melting temperature (~150°C) and that the surface of this protein is almost covered with ion-pair networks.[55] All this evidence supports two basic features of our approach: considering ion pairs as the preferred mode of interaction, and mutation of surface residues to generate more ion pairs for maintaining the protein conformation at extremely high temperatures. It is also important to point out that consideration of surface residues for mutation is consistent with our expectation that mutation of surface residues is not likely to alter the 3D architecture of the protein. Conversely, because of the tight packing in the interior of a protein, mutating any residue there will likely alter the 3D architecture, and hence its functional activity.

Our strategy consists of a set of well-defined steps:

1. We examine the residue interaction pattern of the energy-minimized 3D structure of the original protein in order to identify the polar or charged residues whose side chains interact weakly with the rest of the protein and are considered as the key residues. This interaction energy includes electrostatic and van der Waals components.
2. We consider only such key residues that are close to the surface. Subsequently we identify other nearby residues that interact weakly with the selected key residue. Depending on the situation, these neighboring residues may be polar or nonpolar. The side chains of such residues are chosen for replacement with desired residues in order to improve their interaction with the key residue and hence the stability of the protein.
3. The side chains of each of these selected residues are then replaced by a conformer of the side chain of a residue that produces the best improvement in the interaction energy of it with the rest of the protein. This process is repeated for all the selected residues corresponding to all the selected key residues.
4. In order to refine the structure, the resulting protein is energy minimized using CHARMM (Chemistry at HARvard Macromolecular Mechanics) software considering the full force field.[56,57]
5. The self-energies of the original protein and the mutated protein are computed and compared to demonstrate the substantial gain in stability of the mutated protein. The self-energy (E_{self}) of a protein in its 3D structure represents the total energy content of the structure including all the bonded and nonbonded interaction energy components. The stability of a protein is directly related to the difference between its folded and unfolded forms. However, as in the present case, we are only interested in the difference in stability, we use the difference in self-energies (ΔE_{self}) as a measure of the stability difference between the folded structures of the original and its designed mutant variant. The entire process can be repeated for a few cycles if required to make the stability enhancement more pronounced. The overall computational work flow is summarized in Figure 6.1. The detailed methods of each step are described below.

6.2.2 GENERATION OF SIDE-CHAIN CONFORMER LIBRARIES

In generating the side-chain conformer libraries we randomly downloaded a number of crystal structures of proteins from the Research Collaboratory for Structural Bioinformatics (RCSB) database. From that lot we finally selected 49 Protein Data Bank (PDB) entries that were single-chain globular structures without any missing residues in the respective PDB files. A list of the PDB identification (ID) numbers of these 49 proteins is provided in Table 6.1. First, the H-atoms were assigned to each

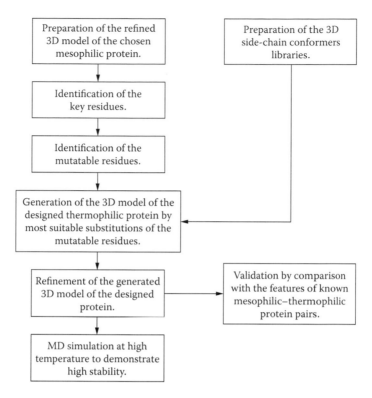

FIGURE 6.1 Schematic diagram summarizing the overall computational work flow of the current design approach.

TABLE 6.1

PDB IDs of the 49 Proteins Used for Generating Conformer Libraries

No.	PDB ID	No.	PDB ID	No.	PDB ID	No.	PDB ID
1	1aky	14	1gro	27	1pst	40	1zin
2	1bjk	15	1gtm	28	1qnm	41	2b3y
3	1bko	16	1guz	29	1qpg	42	2cgg
4	1bpd	17	1h9o	30	1qst	43	2gbl
5	1byc	18	1ho5	31	1rmg	44	2v5l
6	1cdg	19	1hrd	32	1sky	45	2v7y
7	1csh	20	1htb	33	1tim	46	2v6l
8	1dfg	21	1hyg	34	1vis	47	2vj0
9	1dxk	22	1jum	35	1w2c	48	3pfk
10	1eft	23	1lnf	36	1w2d	49	4mdh
11	1egh	24	1mat	37	1w2f		
12	1f8i	25	1mfp	38	1wer		
13	1gr0	26	1nhg	39	1xct		

of these downloaded PDB files. Each H-assigned PDB file was energy minimized in vacuum by 5000 steepest descent steps using the CHARMM energy minimization tool and parameters[56,57] in order to remove bad steric conflicts from the structure. In the energy minimization process we used a distance-dependent dielectric constant and a spherical cutoff of 12.0 Å in computing the nonbonded interactions. All 49 chosen protein structures were processed in this way. Subsequently we extracted the coordinates of the side chains of a particular residue type and stored them in a specific file named by its type, such as Arg.lib. In this way we generated the conformer libraries for each of the 19 (an H-atom being the side chain; glycine was not considered) different residue types. The advantage of generating the side-chain conformer library directly from the crystal structures is that it provides the side-chain conformations that really exist in nature, and it includes the different orientations of the individual conformers as well. We also developed in-house FORTRAN codes for generating the conformer libraries from the PDB files of the proteins.

6.2.3 IDENTIFICATION OF WEAKLY INTERACTING SIDE CHAINS AND THEIR MUTATABLE PARTNER RESIDUES

The 3D structure of the protein under consideration is first refined by minimization of energy following the protocol described in the previous paragraph. Then, considering electrostatic and van der Waals interactions, the energies of interaction of the individual side chains with the rest of the protein are computed using in-house FORTRAN codes. The electrostatic interaction was computed by Coulomb's law using the standard relation

$$E_{elec} = 331.5 \times \sum_{i=1}^{n} \sum_{\substack{j=1 \\ j \neq i}}^{n} \frac{q_i q_j}{\varepsilon \cdot d_{ij}}$$

where q_i represents the partial atomic charge of the ith atom of the respective side chain (excluding the backbone part) of the residue of interest, and q_j represents the same for the atoms of the rest of the protein. We used a distance cutoff value of 12.0 Å and a distance-dependent dielectric constant ($\varepsilon = \varepsilon_0 d_{ij}$), as in CHARMM. The van der Waals interaction was computed using the Lennard–Jones formula:

$$E_{vdw} = \sum_{i=1}^{n} \sum_{\substack{i=1 \\ j \neq i}}^{n} e_{ij} \left[\left(\frac{\sigma_{ij}}{d_{ij}} \right)^{12} - 2 \left(\frac{\sigma_{ij}}{d_{ij}} \right)^{6} \right]$$

where $\sigma_{ij} = (\sigma_i + \sigma_j)/2$, where σ_i and σ_j represent the Lennard–Jones diameters of the ith and jth atoms, respectively, d_{ij} is the interatomic distance between the ith and jth atoms, and $e_{ij} = \sqrt{e_i e_j}$, with e_i and e_j being the Lennard–Jones well depth of the ith and jth atoms, respectively. We used the partial atomic charges and the van der Waals parameters following CHARMM. It should be pointed out here that we used our in-house tool only to identify the key residues and the most suitable mutation in its neighborhood. The final energetics in all cases were computed using CHARMM. In the next step, the generated residue interaction pattern was used to identify the polar or charged residues that interact weakly with the rest of the protein. These residues are considered as the key residues. These weak interactions seem to result from the residue being placed in an electrostatically noncomplementary environment, or there may be a repulsive electrostatic interaction with another similar residue in its vicinity. In cases of the first type, a residue in its immediate vicinity may be identified and subsequently replaced by a polar or charged residue of a complementary nature to gain stability. In the second case, either of the residues may be mutated to achieve better local electrostatic complementarity. It should be emphasized that all these factors depend on the 3D structure of the protein, and contributions come from interactions among many residues that are "sequence-wise local" as well as "nonlocal along the sequence." Only "sequence-wise nonlocal" interactions are expected to be major determinants in stabilization of the 3D structure of a protein. Based on this idea, we select several (approximately 10) such key polar or charged side chains distributed close to the surface over the entire protein. Subsequently we identify potential residues for side-chain substitutions by inspecting the immediate neighborhood of the key residues in the energy-minimized protein structure. If the key residue and its mutatable partner are very close (less than 10 residues) to each other along the sequence, we do not consider that residue, as it is quite unlikely that this will effectively enhance the overall stability of the protein. This criterion eliminates the possibility of false selections due to favorable interactions among sequence-wise local residues, as those favorable interactions are expected to be present in the unfolded forms of the protein, and hence should be ineffective in increasing the stability of the folded protein. After selecting the acceptable mutatable residues, we replace the individual side chains with the best conformer of each of the other residue types obtained by searching the conformer libraries to find the best substitution, as described in the next subsection. It should be emphasized that in our strategy, the mutations are made only to the residues on or very close to the surface of the protein. Hence, it is highly unlikely that the 3D architecture of the protein will be affected significantly due to such mutations. Thus the proteins designed following our strategy are expected to fold and function in a proper way.

6.2.4 Finding Suitable Substitutes and Generation of the 3D Structure of the Designed Protein

We developed in-house FORTRAN codes to identify the best substitution residue side chains by trying all the conformers of the different side chains from the conformer libraries generated earlier. For each type of residue, all the individual side-chain conformers with different spatial orientations are substituted and the interaction energy with the rest of the protein is computed, as explained earlier. The conformer giving the lowest energy of interaction with the rest of the protein is identified as a potential substitution. This is repeated for all 18 (glycine not considered) residues. We select the best replacement considering the most favorable interaction with the rest of the protein and replace the original side chain with it. These steps are repeated to replace all the selected mutatable side chains one by one in a cumulative fashion. In each case we also identify from the rest of the protein the residues contributing most to the favorable interaction with the selected conformer. At the end of the process we find the atomic coordinates of the designed protein with the substituted side chains at the selected residues.

The generated structure of the new mutated protein is then energy minimized following the same protocol as mentioned earlier, followed by computation of the self-energy of the protein using CHARMM.

6.2.5 Demonstration of Enhanced Stability of the Designed Protein

The ideal way to demonstrate the enhancement of stability of the designed protein is to synthesize the designed mutated protein and compare the thermal stabilities of the original mesophilic protein with the designed one by performing melting experiments. However, that is beyond our scope. As an alternative, we have used computational measures for this purpose. Comparison between the self-energies of the energy-minimized original and the mutated proteins is a good way of demonstrating the differences in their thermal stabilities. Self-energy of a protein in its 3D structure represents the total energy content of the structure. The stability of a protein is directly related to the difference between its folded and unfolded forms. However, we are only interested in the difference in stability, and the designed protein differs from the original protein only in the mutations. One can use the difference in self-energies of the thermophilic and mesophilic versions as a measure of the stability difference between the folded structures of the original and its designed mutant variant. Thus the difference $(\Delta E_{self} = E_{self}^{Th} - E_{self}^{Me})$ in the self-energies of the thermophilic (E_{self}^{Th}) and mesophilic versions (E_{self}^{Me}) can be used as a measure of the stability difference between the two. As defined here, a negative value of ΔE_{self} implies that the mutated thermophilic protein is more stable than the corresponding original one. The self-energy of the designed protein is expected to be significantly lower than that of the original protein, indicating improved stability of the designed protein. It should be emphasized that as the protein changes based on its amino acid content and sequence, the difference in the self-energies in the folded state should not strictly represent the difference in stability in an exact way but should be a reasonable indicator of it.

The residue plot of the differences in the interaction energies of the respective individual residues of the protein pair represents the way the change in stability takes place at the individual residue level. We used these methods in some of our earlier works[58–61] as well. It should also be pointed out that since the interresidue interactions are addressed at the atomic level, the applicability of the present method should not have any bias toward any secondary or tertiary structure.

We have also performed MD simulation of the designed protein at a temperature of 350 K (Case Study III) in order to examine whether the designed protein remains stable under dynamic conditions at high temperature and whether the new strong interactions developed due to mutations are maintained or not. MD simulations of the original and mutated proteins were done separately in vacuum using its energy-minimized structures and CHARMM. Newton's equation of motion for each atom was integrated using the leapfrog algorithm[62] with a timestep of 2.0 fsec. The SHAKE algorithm[63,64] was applied to constrain the bond lengths involving H-atoms to their equilibrium positions with a tolerance of 0.0001. Spherical cutoff methods were applied in calculating the nonbonded interactions[65] that were smoothly switched to zero at 11.0 Å. The nonbonded pair list was generated using a cutoff value of 12.0 Å and was updated every 20 steps. For electrostatic calculations, a distance-dependent dielectric constant was used. During the MD simulation, the system was heated to 350 K during the first 10 psec and then equilibrated for 20 psec by assigning velocities to the atoms from a Gaussian distribution at 350 K. The simulation was continued, checking the temperature every 200 steps, and the temperature was adjusted by scaling velocities only if the average temperature of the system was outside the window 350 ± 5 K. Thus, the average temperature was maintained around 350 K. The trajectory was saved every 200 steps for analysis. MD simulation was continued for 4.0 nsec. As a control, similar MD simulation of the original mesophilic protein was also done following identical protocols for comparison.

6.2.6 Validation of the Current Approach

Validation of the methods can be done in a most straightforward way simply by synthesizing the designed mutated protein and comparing the thermal stabilities by performing melting experiments. As an alternative, comparison of the properties of the test proteins with those of known natural mesophilic–thermophilic protein pairs is quite useful.

In order to validate our method of demonstrating stability differences between proteins, we applied it to four widely different known natural mesophilic–thermophilic protein pairs. The individual PDB files were downloaded from the RCSB database. The individual structures were then refined by energy minimization using the same protocol as mentioned earlier. The refined structures were then used to compute the ΔE_{self} values for each protein pair to compare the stabilities of the individual protein pairs. The observed trend of ΔE_{self} values was compared with those of the test proteins. In the cases where the number of residues in the mesophilic protein and its counterpart in a protein pair are different, we use the normalized quantity ΔE_{self} per residue ($\Delta E'_{self}$), defined as

$$\Delta E'_{self} = \frac{E^{therm}_{self}}{N^{therm}} - \frac{E^{meso}_{self}}{N^{meso}}$$

where N^{therm} and N^{meso} represent the number of residues in the thermophilic and mesophilic variants, respectively. Thus $\Delta E'_{self}$ is a useful quantity to compare the relative stability of proteins of different sizes. A larger negative value of $\Delta E'_{self}$ should reflect a larger difference in melting temperatures between the thermophilic and mesophilic protein pair. This exercise validates the use of ΔE_{self} as a good measure of protein structure stability.

The interaction energies of individual residues with the rest of the protein were also computed. We then aligned the two sequences of the thermophilic–mesophilic protein pair in order to identify the respective residues in the two proteins based on one-to-one mapping of the aligned residues. Then we removed the gaps by deleting the residues from the other protein and computed the ΔE_i values for the aligned respective residue pairs from the two protein sequences. The observed patterns of the differences in the residue interaction energies were then compared to those of the test proteins in order to identify pattern similarities. This study validates the use of interaction patterns in identifying mutating residues and side-chain substitution as a method of generating thermophilic proteins.

Finally, we used our proposed methods to generate a thermophilic counterpart for each mesophilic protein for a set of known mesophilic–thermophilic protein pairs to show that many of the suggested mutations are consistent with the natural thermophilic variants. This provides overall validation for our approach.

6.3 RESULTS

6.3.1 GENERATION OF SIDE-CHAIN CONFORMER LIBRARIES

The conformer libraries for all the different kinds of amino acid residues (excepting glycine) were generated using our in-house tool and 49 PDB files as explained in Section 6.2. The generated conformers were visually inspected and were found to represent the conformational space available to the side chain reasonably well. The conformer models also cover their diverse orientations satisfactorily. The conformer libraries generated and used here are reasonably large. The PDB IDs of the proteins we used for generating the conformer libraries are provided in Table 6.1.

6.3.2 SELECTION OF TEST CASES

In order to demonstrate our design strategy of converting a mesophilic protein into a thermostable one, we selected the crystal structures of three moderate-size single-chain globular proteins without any missing coordinates of the intermediate residues. These representative mesophilic proteins are rat intestinal fatty acid-binding protein (PDB ID 2ifb), sialic acid-binding protein (SIAP) from *Haemophilus influenzae* (PDB ID 2v4c), and phosphinothricin acetyltransferase from *Pseudomonas aeruginosa* (PDB ID 2bll). The PDB files representing the crystallographic 3D structures were collected from the RCSB database and H-atoms were assigned to each protein. The resulting models were then refined by energy minimization as described in Section 6.2.

6.3.2.1 Case Study I: Rat Intestinal Fatty Acid-Binding Protein (PDB ID 2ifb)

6.3.2.1.1   Identification of Weakly Interacting Side Chains and Their Mutatable Partner Residues

The side-chain interaction (interaction energies of the side chains of the individual residues with the rest of the protein) pattern for the energy-minimized protein (PDB ID 2ifb) was generated using our in-house FORTRAN codes, as explained in Section 6.2. Analysis of the residue interaction pattern (Figure 6.2) revealed that

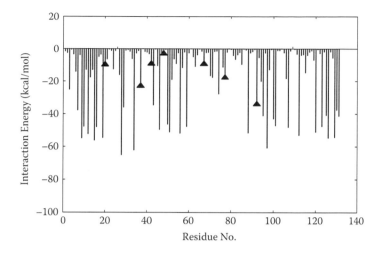

FIGURE 6.2 Interaction pattern of the individual residues against the residue number for the protein crystal structure 2ifb. The residues selected as the key residues are indicated by a filled triangle at the bottom of the respective bar; all these residues exhibit poor interaction energies with the rest of the protein.

several polar/charged residues interacted with the rest of the protein much weaker compared to the same residue types at other places. This implies that they do not have an electrostatically complementary residue in their immediate neighborhood. Of these residues, we selected Lys[20], Lys[37], Gln[42], Thr[48], Asp[67], Glu[77], and Lys[92] as the key residues whose interactions with the rest of the protein need to be enhanced in order to enhance the overall stability of the protein. During selection, care was taken to ensure that the selected residues were well distributed over the protein surface and not localized in a small zone. We then inspected the neighborhood of each such key residue and identified a best-suited nearby residue that could be replaced by a residue of complementary nature to enhance the local interaction. Based on these key residues, the respective mutatable residues chosen were Gly[121], Ser[52], Met[84], Val[61], Thr[81], Val[96], and Ile[103]. The key residues, their respective mutatable residues selected in this way, the spatial distance (D_{CA}) between the CA atoms of the key residue and its mutatable partner, and the number of residues Δres separating the key residue and the respective mutatable residue along the sequence are summarized in Table 6.2. It should be pointed out that in none of the cases, the key residue and its mutatable partner are closer than 10 residues along the sequence, even though they are spatially close enough to interact with each other. Thus the mutations should enhance the overall stability of the 3D architecture by stabilizing sequentially distant but spatially local parts of the protein.

6.3.2.1.2 Generating the Best Mutated Protein

For each selected mutatable residue mentioned in the test protein structure 2ifb, the respective side chain was replaced sequentially by the individual conformers of the side chains of the other residues from the pregenerated conformer libraries and for each substitution its interaction energy with the rest of the protein was computed by

TABLE 6.2

Summary of the Selected Key Residues, Identified Mutatable Residues in its Vicinity, the Best Mutation, and the Respective Computed Energy Difference

PDB ID	Sequence Identity	No.	Key Residue	Mutatable Residue	Δres	Mutation Made	ΔE_i (kcal/mol)	D_{CA} (Å)
2ifb	94.7%	1	Lys[20]	Gly[121]	101	Glu[121]	−38.9	8.8
		2	Lys[37]	Ser[52]	15	Asp[52]	−29.8	5.3
		3	Gln[42]	Met[84]	42	Arg[84]	−26.1	11.3
		4	Thr[48]	Val[61]	13	Arg[61]	−50.4	4.6
		5	Asp[67]	Thr[81]	14	Arg[81]	−63.4	4.4
		6	Glu[77]	Val[96]	19	Lys[96]	−47.2	5.5
		7	Lys[92]	Ile[103]	11	Glu[103]	−43.4	4.9
2v4c	97.7%	1	Asp[4]	Ser[39]	35	Arg[39]	−80.4	7.6
		2	Arg[50]	Asn[150]	100	Glu[150]	−49.8	8.8
		3	Asp[113]	Tyr[75]	38	Lys[75]	−14.0	10.5
		4	Arg[133]	Leu[144]	11	Glu[144]	−26.0	8.6
		5	Asn[136]	Thr[268]	132	Glu[268]	−22.6	6.5
		6	Glu[225]	Tyr[3]	222	Lys[3]	−46.3	10.3
		7	Lys[276]	Tyr[85]	191	Glu[85]	−35.4	6.8
2bl1	95.9%	1	Glu[33]	Phe[152]	119	Arg[152]	−19.4	8.0
		2	Gln[42]	Val[10]	32	Lys[10]	−52.7	7.9
		3	Arg[75]	Phe[160]	85	Asp[160]	−28.3	13.4
		4	Gln[102]	Ile[5]	97	Arg[5]	−63.4	7.7
		5	Arg[112]	Leu[163]	51	Asp[163]	−13.7	10.9
		6	Arg[136]	Val[101]	35	Glu[101]	−36.1	9.6
		7	Asp[157]	Pro[146]	11	Arg[146]	−46.4	4.5

using our in-house FORTRAN codes. Based on the interaction energies, the best substitution was selected. Table 6.2 summarizes the suggested substitution and the resulting gain in stability. It should be noted that all the best substituted side chains obtained are charged ones, such as Arg, Lys, Glu, and Asp. Table 6.2 clearly demonstrates that the suggested substitution of the side chain in each case resulted in significant improvement in its interaction energy with the rest of the protein, as represented by $\Delta E_i = E_i^{Th} - E_i^{Me}$, where E_i^{Th} represents the interaction energy of the ith residue with the rest of the mutated protein and E_i^{Me} represents the same for the original mesophilic protein. Improvements in the interaction energies due to side-chain replacement were found to be in the range −26.1 kcal/mol to −63.4 kcal/mol. The atomic coordinates of the mutated side chains were directly generated by our in-house codes from the conformer libraries. For each mutatable residue, the atomic coordinates of the mutated side chain obtained in the previous step were substituted to the side chain of the respective mutatable residue. The mutated protein obtained in this way was then refined by energy minimization by 5000 steps of the steepest descent algorithm using CHARMM in order to better accommodate the substituted side chain into the protein.

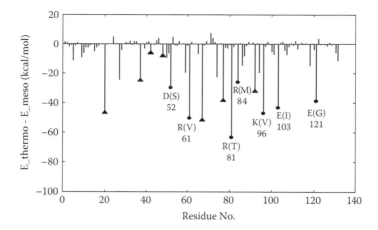

FIGURE 6.3 The pattern of the differences $(\Delta E_i = E_i^{Th} - E_i^{Me})$ in the residue interaction energies between the original protein and the designed mutated protein in the case of the crystal structure 2ifb. The key residues and the mutated residues are highlighted by filled triangles and filled circles, respectively.

6.3.2.1.3 Demonstration of Enhanced Stability of the Designed Protein

Figure 6.3 compares the profile of the residue interaction energy difference $(\Delta E_i = E_i^{Th} - E_i^{Me})$ of the original and the mutated proteins. It is clear that the mutatable residues that showed weaker interactions with the rest of the protein in the original protein are now after mutations showing stronger interactions in the corresponding mutated protein, as expected in the design strategy. Moreover, Figure 6.3 shows that, consistent with the design concepts, the most favorably interacting partners of the mutated residues are the respective key residues.

It should also be noted that not only the interaction energies of the mutated residues and the key residues are improved, but the interaction energies of several other residues are also significantly improved due to the mutations (Figure 6.3). Such residues are Lys[27], Asp[59], Asp[74], Lys[94], Thr[118], and Glu[131], whose interactions with the mutated residues increased the overall stability of the protein.

Table 6.3 summarizes the comparison of the differences in the self-energies (ΔE_{self}) between the mutated and original proteins. It can be clearly seen that the

---

TABLE 6.3

Comparison of the Self-Energies between the Mesophilic and Thermophilic Versions of the Three Test Proteins

Protein Name	No. of Residues	E_{self}^{Me} (kcal/mol)	E_{self}^{Me} (kcal/mol)	ΔE_{self} (kcal/mol)	$\Delta E'_{self}$ (kcal/mol)
2ifb	131	−1615.2	−2162.2	−547.0	−4.18
2v4c	309	−2747.3	−3130.6	−383.3	−1.24
2bl1	172	−2013.0	−2523.6	−510.6	−2.97

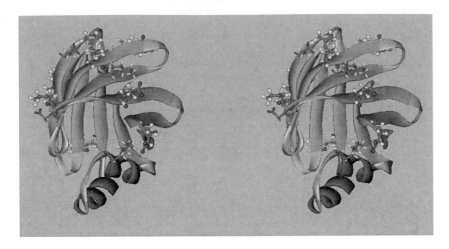

FIGURE 6.4 Stereo view of the 3D structure of the designed protein for the case of the crystal structure 2ifb with the key residues and mutated residues highlighted as sticks and balls and sticks, respectively. **(See color insert.)**

designed mutant variant is significantly more stable (ΔE_{self} = −547.0 kcal/mol) than the corresponding original mesophilic protein.

Comparison of the $\Delta E'_{self}$ value of the mutated protein 2ifb with that of the known mesophilic–thermophilic protein pair 1csp-1c9o further indicates that the enhancement in stability of this mutated protein should be much higher compared to that ($\Delta E'_{self}$ = −2.0 kcal /mol) of the known mesophilic–thermophilic pair 1csp-1c9o. Thus the designed protein in this case study seems to be a hyperthermophilic one.

Figure 6.4 represents the stereo view of the 3D structure of the energy-minimized mutated variant for the test protein with the key residues and mutated side chains highlighted. It is clear that all the mutated residues are close to the surface of the protein, as originally planned, and they are well distributed over the protein's surface. Thus it appears that a number of local stabilizations enhanced the global stability of the protein.

6.3.2.2 Case Study II: Sialic Acid-Binding Protein (SIAP) from *Haemophilus influenzae* (PDB ID 2v4c)

6.3.2.2.1 *Identification of Weakly Interacting Side Chains and Their Mutatable Partner Residues*

The residue interaction pattern in the original protein is shown in Figure 6.5. Calculations and analysis similar to Case Study I were also done for this test case protein (2v4c) and the results are summarized in Table 6.2. In this case, based on the residue interaction energy profile, the seven residues—Asp[4], Arg[50], Asp[113], Arg[133], Asn[136], Glu[225], and Lys[276]—were selected as the key residues. Then, in order to improve their contribution in the self-energy of the protein for mutations we selected the residues Ser[39], Asn[150], Tyr[75], Leu[144], Thr[268], Tyr[3], and Tyr[85], respectively, in their neighborhood. The Δres values in Table 6.2 clearly indicate that none of these selected key residues and their respective mutatable partners are "sequence-wise close," and their D_{CA} values ensured that they are "spatially close" enough for strong interactions.

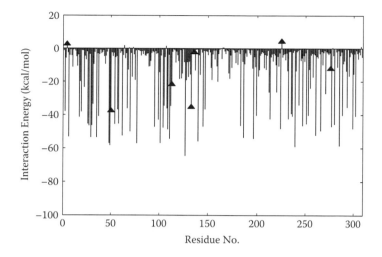

FIGURE 6.5 Pattern of the interaction energies of the individual residues with the rest of the protein versus the residue number for the protein crystal structure 2v4c. The residues selected as the key residues are indicated by a filled triangle at the bottom of the respective bar; all these residues exhibit poor interaction energies with the rest of the protein.

6.3.2.2.2 Generating the Best Mutated Protein

We followed the same procedures as mentioned in Section 6.2 for generating the best side-chain replacements for the mutatable residues in the test protein 2v4c. The results are summarized in Table 6.2, and it is quite apparent that all the individual mutations improved their contributions to the self-energy of the protein and thus should further stabilize the protein. The gain in interaction energy in the individual residue level is found to be in the range −14.0 kcal/mol to −80.4 kcal/mol. It is quite interesting that here the most suitable substituted side chains are found to be charged side chains like Glu, Lys, and Arg.

6.3.2.2.3 Demonstration of Enhanced Stability of the Designed Protein

Figure 6.6 represents the ΔE_i versus the residue number for the protein 2v4c. Here also, the growth of several strong and favorable interactions in the mutated protein is clearly seen, as was found in Case Study I. It was found that all the rationally mutated residues contributed significantly in improving the overall stability of the mutated protein. However, the most effective contributions came from the residues Arg[39], Lys[3], and Glu[105]. There are two key residues, Arg[133] and Asn[136], whose interactions with the rest of the protein did not improve much compared with the others.

Table 6.3 indicates that the mutated protein is significantly stable compared to the original protein (2v4c), with $\Delta E_{self} = -383.3$ kcal/mol. Comparison between the computed values of ΔE_{self} per residue ($\Delta E'_{self}$) for the proteins in Case Study I and Case Study II further indicates that the degree of enhancement of stability for the protein in Case Study II is much less compared with the protein in Case Study I. Similar comparison of the $\Delta E'_{self}$ value of the mutated 2v4c protein with that ($\Delta E'_{self} = -2.0$ kcal/mol) of the known mesophilic–thermophilic pair 1csp-1c9o indicates that the

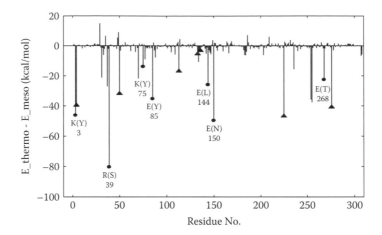

FIGURE 6.6 The pattern of the differences ($\Delta E_i = E_i^{Th} - E_i^{Me}$) in the residue interaction energies between the original protein and the designed protein in the case of the crystal structure 2v4c. The key residues and the mutated residues are highlighted by filled triangles and filled circles, respectively.

enhancement in stability of this mutated protein should be less compared to that of the 1csp-1c9o pair.

Like Case Study I, here also a number of residues other than the key residues and mutated ones are found to contribute in significantly improving the stability of the protein through interactions with the mutated residues. These residues are Lys[31], Glu[37], Arg[70], Ala[134], Asp[236], Lys[254], and Asp[255]. It must be pointed out that the two residues Lys[254] and Asp[255] are consecutive along the sequence, and the improvement in interactions they showed was due to their mutual interactions and hence is not at all useful in stabilizing the folded structure of the protein.

Figure 6.7 represents the stereo view of the 3D structure of the energy-minimized mutated variant for the test protein with the key residues and mutated side chains highlighted. Note that all the mutated residues are close to the surface of the protein and are well distributed over the protein's surface.

6.3.2.3 Case Study III: Phosphinothricin Acetyltransferase from *Pseudomonas aeruginosa* (PDB ID 2bl1)

6.3.2.3.1 Identification of Weakly Interacting Side Chains and Their Mutatable Partner Residues

Here also, calculations and analysis similar to Case Study I were made for the test protein (2bl1) and the results are summarized in Table 6.2.

In this case, based on the residue interaction profile (Figure 6.8), the residues Glu[33], Gln[42], Arg[75], Gln[102], Arg[112], Arg[136], and Asp[157] were identified as good key residues whose interactions with the rest of the protein needed to be improved by mutations in their immediate neighborhood. The residues Phe[152], Val[10], Phe[160], Ile[5], Leu[163], Val[101], and Pro[146] were subsequently identified as the most suitable mutatable residues, respectively. The Δ*res* values in Table 6.2 confirm that the key residues

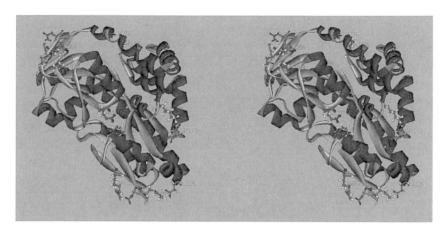

FIGURE 6.7 Stereo view of the 3D structure of the designed protein for the case of the crystal structure 2v4c with the key residues and mutated residues highlighted as sticks and balls and sticks, respectively. **(See color insert.)**

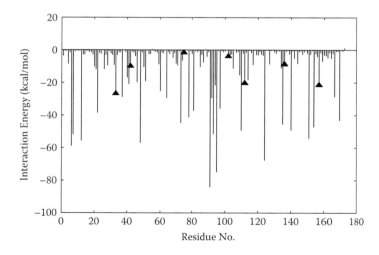

FIGURE 6.8 Pattern of the interaction energies of the individual residues with the rest of the protein against the residue number for the protein crystal structure 2bl1. The residues selected as the key residues are indicated by a filled triangle at the bottom of the respective bar; all these residues exhibit poor interaction energies with the rest of the protein.

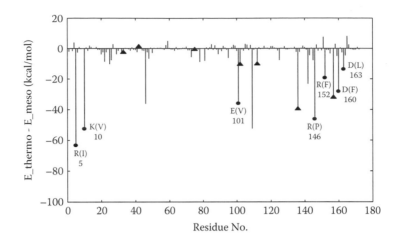

FIGURE 6.9   The pattern of the differences $(\Delta E_i = E_i^{Th} - E_i^{Me})$ in the residue interaction energies between the original protein and the designed protein in the case of the crystal structure 2bl1. The key residues and the mutated residues are highlighted by filled triangles and filled circles, respectively.

and their mutatable partners are sufficiently distant in the sequence, and D_{CA} values ensure that they are spatially close and thus satisfy our selection criteria.

6.3.2.3.2 Generating the Best Mutated Protein

Following the methods of substitution of the side chains of the mutatable residues, as in the previous two case studies, the best mutations we identified were Arg[152], Lys[10], Asp[160], Arg[5], Asp[163], Glu[101], and Arg[146]. Figure 6.9 represents the ΔE_i versus the residue number plot for the current test protein, 2bl1. Here also, the growth of several strong and favorable interactions in the mutated protein is clearly demonstrated. It was found that all the mutated residues contributed significantly in enhancing stability.

As in Case Study I, all the individual mutations improved their interactions with the rest of the proteins, thus improving the stability of the protein. The gain in interaction energy at the individual level was found to be in the range −13.7 kcal/mol to −63.4 kcal/mol (Table 6.2). The most suitable substituted side chains were again found to be Asp, Glu, Lys, and Arg, all of which have charged side chains.

Two residues, Arg[152] and Asp[163], were found to be relatively weak compared with the other mutated residues. The most effective contributions to the stability of the mutated protein came from the residues Arg[5] and Lys[10]. Apart from these mutated residues, several other residues, including Gly[25], Asp[46], Glu[109], and Ser[142], which were not mutated, but gained significantly in stabilizing the resulting protein, mostly due to their interactions with the mutated residues.

Stereo views of the 3D structure of the energy-minimized mutated variant of the test protein, with the key residues and mutated side chains highlighted, are presented in Figure 6.10. As in the other cases, all the mutated residues were found to be close to the surface of the protein and are well distributed over the protein's surface. Thus,

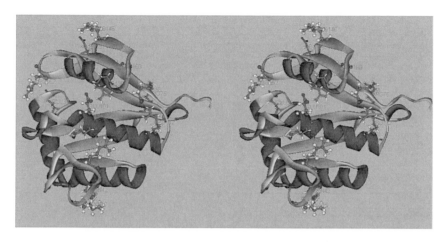

FIGURE 6.10 Stereo view of the 3D structure of the designed protein for the case of the crystal structure 2bll with the key residues and mutated residues highlighted as sticks and balls and sticks, respectively. **(See color insert.)**

the enhanced global stability of the mutated protein is driven by a number of spatially local stabilizations.

6.3.2.3.3 Demonstration of Enhanced Stability of the Designed Protein

Table 6.3 clearly indicates that the mutated protein is significantly stable compared to the original protein 2bll, with $\Delta E_{\text{self}} = -510.6$ kcal/mol. Comparison between the computed $\Delta E'_{\text{self}}$ values for the original and mutated 2bll with reference to the protein pairs in Case Study I and Case Study II implies that the degree of enhancement of stability for the designed protein in Case Study III is greater compared to that of the designed protein in Case Study II, but less compared to that in Case Study I. Further comparison of the $\Delta E'_{\text{self}}$ value of the present protein pair with that of the known natural mesophilic–thermophilic protein pair (1csp-1c9o) further indicates that the enhancement in stability of this mutated protein should be comparable to $(\Delta E'_{\text{self}} = -2.0$ kcal/mol) the known mesophilic–thermophilic pair 1csp-1c9o.

Figure 6.9 further indicates that the residues Gly[25], Asp[46], Glu[109], and Ser[142] have improved their interactions with the rest of the protein in spite of the fact that they are not the key residues or the mutated residues.

Figure 6.11 shows that over the 4.0 nsec MD simulation in a vacuum, the mesophilic protein deviated much more from its initial structure than its designed thermophilic (mutated) counterpart. The trajectory of the mesophilic protein is showing a clear trend of further deviation in structure, while the thermophilic protein represents a more stable trajectory. This indicates that the mutations have not destabilized the designed protein. Furthermore, comparison of the time evolutions of the self-energies of the original and mutated proteins (Figure 6.12) demonstrated that the designed protein retained its enhanced stability throughout the simulation. The computed values of the self-energies averaged over the last 200 psec of the respective trajectories of the original and mutated proteins are −371.5 kcal/mol and −803.2 kcal/mol, respectively. Thus the designed protein is found to be more stable

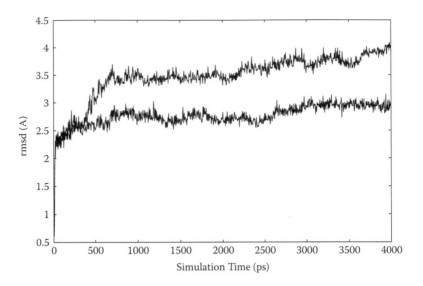

FIGURE 6.11   Comparison of the time evolution of the root mean squared deviation (RMSD) values of 2bl1 and its designed mutated variant with reference to the respective energy-minimized initial structures over the 4.0 nsec trajectory of the MD simulations in vacuum.

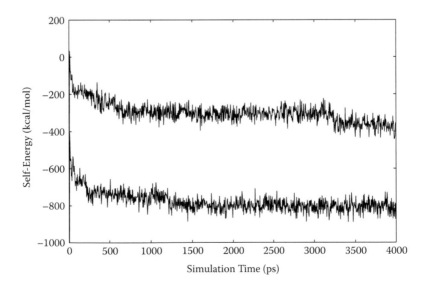

FIGURE 6.12   Comparison of the time evolution of the self-energy of 2bl1 and its designed variant over the 4.0 nsec trajectory of the MD simulations in vacuum.

by −431.7 kcal/mol compared with the original protein, even under dynamic conditions. Figure 6.13 summarizes the time evolutions of the interaction energies of the individual mutated residues with the rest of the protein. It was found that in all the cases the selected mutations maintained their enhanced interaction energies, with some modulations in a few cases. Thus the designed interactions at the individual mutated residues level are well maintained over the entire period of simulation.

We also performed MD simulations of the mesophilic–thermophilic protein pair (1c9o, 1csp) following the same protocol and observed dynamic features as in Case Study III (data not shown). The enhanced stability was reflected in the time evolution of self-energies of the protein pair, which is similar to that in Case Study III (Figure 6.12). The time evolutions of the interaction energies of mutated residues at the individual level were also found to behave in a similar way (data not shown). However, for a more robust study and a realistic comparison, one needs to perform the MD simulation in aqueous solvent.

6.4 VALIDATION OF THE CURRENT DESIGN STRATEGY IN IMPROVING STABILITY

In order to validate our design strategies, we first tried to establish a qualitative correlation between the ΔE_{self} value and the relative stability of a known natural mesophilic–thermophilic protein pair. We considered four known natural thermophilic–mesophilic protein pairs (inorganic pyrophosphatase: 2prd, 1ino; oxidoreductase: 3mds, 1qnm; rubredoxin: 1caa, 8rxn; and cold shock protein: 1c9o, 1csp) for this validation exercise. After downloading crystal structures from the RCSB database, H-atoms were assigned to each of the above-mentioned crystal structures followed by energy minimization, as mentioned earlier. The self-energy of each energy-minimized crystal structure was then computed and subsequently the ΔE_{self} value for each mesophilic–thermophilic protein pair was evaluated. Table 6.4 summarizes different features of the four mesophilic–thermophilic protein pairs. The computed ΔE_{self} values were found to be large and negative, indicating a considerable increase in the stability of the respective thermophilic variant compared with its mesophilic counterpart. These data directly validate the correlation between the ΔE_{self} value and the relative stabilities of the protein pairs, at least qualitatively.

Among the four protein pairs chosen here for validation, 1c9o and 1csp represent experimentally the most well-studied protein pair,[10] and are of almost equal lengths. There are 12 differences in the amino acid residues in the sequences between the thermophilic variant (1c9o) and its mesophilic counterpart (1csp). Table 6.4 shows that the computed ΔE_{self} and $\Delta E'_{self}$ values are −123.9 kcal/mol and −2.0 kcal/mol, respectively, indicating a significant increase in the stability of the thermophilic variant 1c9o. This is quite consistent with the fact that the melting temperature of 1c9o is 23°C higher than that of 1csp.[10] Moreover, comparison of the ΔE_{self} and $\Delta E'_{self}$ values for the other protein pairs clearly indicates that, excepting the case of (1qnm-3mds) pair, the respective thermophilic proteins are not only thermally more stable than their mesophilic counterparts, but are more stable than 1c9o. The natural thermophilic counterpart of 1qnm seems to be less stable than 1c9o.

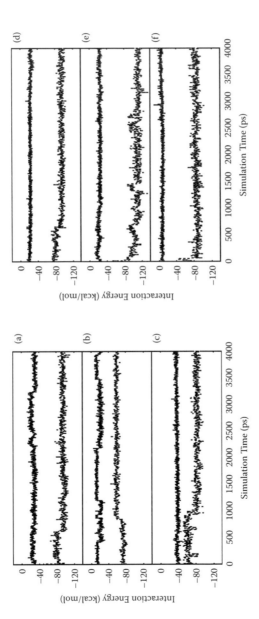

FIGURE 6.13 Time evolutions of the interaction energies of a few individual mutated residues with the rest of the protein over the 4.0 nsec trajectory of the MD simulations in vacuum.

TABLE 6.4

Comparison of the Self-Energies between the Known Natural Mesophilic–Thermophilic Protein Pairs

Protein Pair	Thermo Meso	Sequence Identity (%)	E_{self}^{Me} (kcal/mol)	E_{self}^{Me} (kcal/mol)	ΔE_{self} (kcal/mol)	$\Delta E'_{self}$ (kcal/mol)
2prd (174)	1ino (175)	47.4	−1224.0	−1816.7	−592.7	−3.4
3mds (203)	1qnm (198)	51.3	−1478.1	−1714.9	−236.8	−1.0
1caa (53)	8rxn (52)	70.8	−93.6	−272.5	−178.9	−3.3
1c9o (66)	1csp (67)	84.6	−437.1	−561.0	−123.9	−2.0

It is interesting to note that even though the number of residues in the two proteins in each protein pair is very similar, the sequence identity is quite poor in general (Table 6.4). This simply means that there is a large number of residue mismatches between the proteins in each pair. It should be emphasized here that all the mismatched residues between the thermophilic–mesophilic protein pair are not expected to be responsible for the observed thermostability, and most likely such mismatches have an evolutionary origin, as these are naturally occurring thermophilic–mesophilic protein pairs. Thus the proteins in natural mesophilic–thermophilic pairs are, sequence-wise, more diverse than the designed ones, as those are designed starting with the mesophilic counterpart.

Figure 6.14 shows the residue interaction patterns in the mesophilic protein of each of the four known natural protein pairs. The mutated residues in the respective natural thermophilic protein are mapped onto the mesophilic counterpart and are marked individually. It can be seen that in all the cases, such residues are associated with weak interactions with the rest of the mesophilic protein. We further plotted ΔE_i against i (Figure 6.15) for the four protein pairs and observed similar features as in the three case studies (Figure 6.3, Figure 6.6, and Figure 6.9). All these justify our rationale of identifying the key residues and the mutatable residues in a general way. However, it should be pointed out here that since the sequence identities between the thermophilic and mesophilic counterparts were not very high for the protein pairs, significant noise was observed in the ΔE_i against i plots arising mainly from the nonconserved residues between the protein pairs.

Finally, in order to further validate our approaches, we compared the thermophilic proteins designed by our present method to the respective natural thermophilic proteins. All four known natural mesophilic–thermophilic protein pairs were used for this purpose. In each case, starting with the mesophilic protein, we generated a thermophilic counterpart following our method. It should be noted that only the mismatched residues that were nonpolar in the mesophilic protein and were mutated to a polar or charged residue in the natural thermophilic protein were considered for mutations, as we use electrostatic and van der Waals interactions for finding the best mutation. Table 6.5 and Table 6.6 summarize the results and compare the natural thermophilic protein with the designed protein by mutating the respective mesophilic protein using our methods.

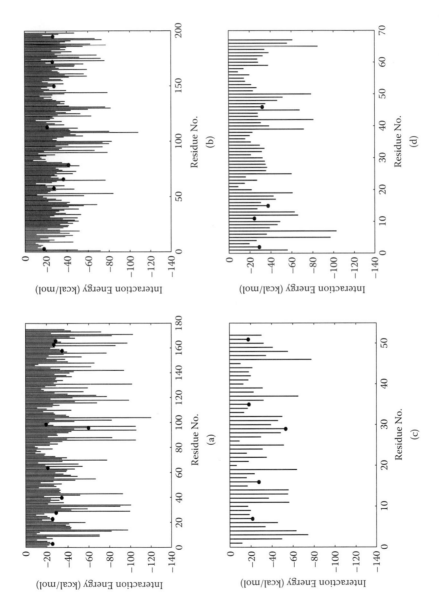

FIGURE 6.14 The residue interaction pattern (E_i vs. i) in the natural mesophilic protein for (a) 1ino, (b) 1qnm, (c) 8rxn, and (d) 1csp. In each case, the mutable residues are marked with a filled circle at the bottom of the respective bar.

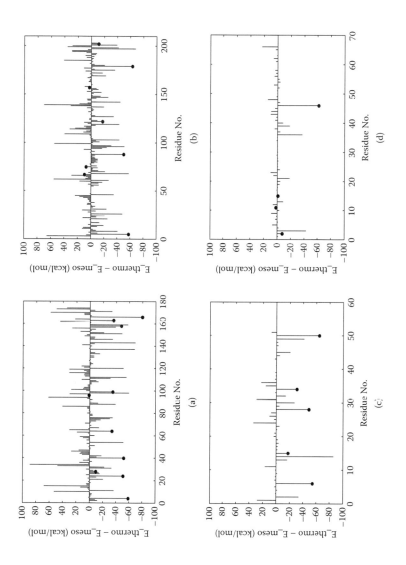

FIGURE 6.15 The pattern of the differences ($\Delta E_i = E_i^{Th} - E_i^{Me}$) in the residue interaction energies between the natural mesophilic protein and its natural thermophilic counterpart for (a) 2prd, (b) 3mds, (c) 1caa, and (d) 1c9o. The mutated residues are highlighted by a filled circle. In each case, the residues that are mutated (nonpolar → polar/charged) in the respective natural thermophilic counterpart are marked with a filled circle at the bottom of the respective bar.

TABLE 6.5
Validation Data for the Natural Thermophiles

Mesophilic Protein Name	Mutatable Residue	Residue in Natural Thermophile	Best Interacting Partner	Δres	ΔE_i (kcal/mol)	ΔE_{self} (kcal/mol)	$\Delta E'_{self}$ (kcal/mol)
1ino	Leu^3	Lys^4	Asp^{67}	63	−59.2	−592.7	−3.4
	Ala^{23}	Arg^{24}	Ser^{75}, Thr^{76} bb*	51, 52	−51.3		
	Ile^{28}	Asn^{28}	Tyr^{30}	2	−9.7		
	Phe^{40}	Lys^{40}	Glu^{31}	9	−52.3		
	Leu^{64}	Glu^{64}	Arg^{171}	107	−34.3		
	Thr^{96}	Glu^{96}	Thr^{153}	57	0.7		
	Ala^{99}	Lys^{99}	Asp^{97}, Glu^{98}	1, 2	−35.4		
	Ala^{158}	Arg^{158}	Asp^{120}	38	−48.1		
	Ala^{163}	Glu^{163}	Arg^{166}	3	−36.2		
	Val^{166}	Arg^{166}	Glu^{14}, Glu^{163}	152, 3	−80.4		
1qnm	Ser^3	Lys^5	Asp^8	3	−58.2	−236.8	−1.0
	Ile^{58}	Gln^{67}	Asp^{68}	1	9.5		
	Phe^{66}	Asn^{75}	—	—	7.1		
	Thr^{79}	Arg^{88}	Glu^{198}	110	−50.4		
	Ala^{113}	Gln^{122}	Glu^{118}	4	−17.9		
	Thr^{150}	Glu^{157}	—	—	2.6		
	Val^{172}	Arg^{179}	Asp^{182}, Gln^{177} bb*	3, 2	−63.6		
	Ala^{195}	Lys^{202}	Glu^{56}	46	−11.7		
8rxn	Thr^7	Lys^6	Glu^{49}	43	−54.7	−178.9	−3.3
	Ala^{16}	Asp^{15}	Lys^2	13	−18.1		
	Ser^{29}	Lys^{28}	Glu^{14}, Glu^{30}	14, 2	−49.1		
	Ala^{35}	Asp^{34}	Lys^{45}	11	−31.1		
	Ala^{51}	Lys^{50}	Glu^{30}, Glu^{52}	20, 2	−64.8		
1csp	Leu^2	Gln^2	Arg^3	1	−7.9	−123.9	−2.0
	Ser^{11}	Asn^{11}	Thr^{40}	29	1.4		
	Phe^{15}	Tyr^{15}	Asn^{10}	5	−1.4		
	Ala^{46}	Glu^{46}	Lys^5	41	−62.1		

Note: bb* = backbone.

Table 6.5 and Table 6.6 indicate that in the case of designing the thermophilic protein starting with the mesophilic counterpart (1ino), of the 10 mutations found in the natural thermophilic protein, for mutations at four positions (Leu^3, Ala^{23}, Ala^{99}, Val^{166}), our best substitutions completely matched with mutated residues found in the natural thermophilic protein. In two cases (Phe^{40}, Ala^{158}), the electrostatic nature of the mutations was found to be the same, even though the residue identities were different. In the four other positions (Ile^{28}, Leu^{64}, Thr^{96}, Ala^{163}), our best substitutions did not match with the respective residues in the natural thermophilic protein. This happened because we identified the best substitutions based on the electrostatic and van der Waals interactions of the side chain with the rest of the mesophilic protein. It

TABLE 6.6

Validation Data for Thermophiles Designed by Our Method

Mesophilic Protein Name	Mutatable Residue	Best Mutation by Our Protocol	Best Interacting Partner	Δres	ΔE_i (kcal/mol)	ΔE_{self} (kcal/mol)	$\Delta E'_{self}$ (kcal/mol)
1ino	Leu^3	Lys^3	Asp^{33}	30	−43.4	−873.9	−5.0
	Ala^{23}	Arg^{23}	Asn^{24}, Thr^{75}	1, 52	−52.2		
	Ile^{28}	Arg^{28}	Asp^{26}	2	−58.2		
	Phe^{40}	Arg^{40}	Glu^{31}, Asp^{42}	9, 2	−71.5		
	Leu^{64}	Arg^{64}	Glu^{101}, Glu^{164}	37, 100	−54.5		
	Thr^{96}	Arg^{96}	Glu^{101}, Glu^{153}	5, 57	−58.2		
	Ala^{99}	Lys^{99}	Asp^{65}, Asp^{102}	34, 3	−57.3		
	Ala^{158}	Lys^{158}	Asp^{122}	36	−34.3		
	Ala^{163}	Lys^{163}	Glu^{159}	4	−49.9		
	Val^{166}	Arg^{166}	Glu^{170}	4	−30.4		
1qnm	Ser^3	Arg^3	Asp^6	3	−67.2	−666.9	−3.4
	Ile^{58}	Arg^{58}	Thr^{55}	3	−19.2		
	Phe^{66}	Glu^{66}	Asn^{142}, Asn^{143}	76, 77	−39.9		
	Thr^{79}	Arg^{79}	Glu^{191}	112	−14.0		
	Ala^{113}	Glu^{113}	Lys^{110}	3	−36.5		
	Thr^{150}	Arg^{150}	Ala^{139} bb*	11	−19.1		
	Val^{172}	Lys^{172}	Asp^{175}	3	−33.3		
	Ala^{195}	Arg^{195}	Glu^{191}	4	−41.0		
8rxn	Thr^7	Lys^7	Glu^{50}	43	−52.2	−337.0	6.5
	Ala^{16}	Arg^{16}	Asp^{14}, Glu^{17}	2, 1	−81.6		
	Ser^{29}	Lys^{29}	Asp^{32}	3	−29.1		
	Ala^{35}	Arg^{35}	Asp^{36}	1	−40.3		
	Ala^{51}	Glu^{51}	Lys^2	49	−31.9		
1csp	Leu^2	Arg^2	Glu^{21}	19	−62.0	−256.7	−3.8
	Ser^{11}	Arg^{11}	Glu^{43}	32	−60.4		
	Phe^{15}	Lys^{15}	Lys^{13} bb*, His^{29}	2, 14	−24.8		
	Ala^{46}	Lys^{46}	Glu^3	43	−52.8		

Note: bb* = backbone.

must be emphasized here that the natural thermophilic protein is not necessarily the best thermostable variant. Comparison of the ΔE_{self} and $\Delta E'_{self}$ values for the natural thermophilic protein and our designed protein, as given in Table 6.5 and Table 6.6, clearly indicates that our designed protein is much more stable compared with the natural thermophilic protein. For individual mutations, we examined the interaction energy of the mutated residue with the rest of the protein when the mutation in 1ino was made by the residue type found in the respective thermophilic protein (2prd) using our method. This interaction energy was found to be favorable compared with that in the mesophilic counterpart (1ino), but not to the extent we found in our design. The identified best partner residues were also different in the two cases. The natural selection of mutation may be influenced by other factors as well. Thus, the naturally occurring mutation improves the stability, but not in the best possible way. Moreover, our design is based on the mesophilic protein; the designed thermophilic protein differs only at the mutated residues, while the natural thermophilic protein differs in other places, and that may influence the nature of the mutations.

Analysis indicates similar features for the other three known mesophilic–thermophilic protein pairs. In the case of 1qnm, out of eight mutations, only one mutation at position Thr[79] was found to be identical between the natural protein and our designed protein. For five mutations at positions Ser[3], Phe[66], Ala[113], Val[172], and Ala[195], the electrostatic nature was preserved, and in two positions (Ile[58], Thr[150]) the electrostatic nature of the mutation was opposite (the possible reasons are discussed in the previous paragraph).

In the case of 8rxn, out of five mutations, complete agreement was obtained for two residues at positions Thr[7] and Ser[29]. For the other three cases (Ala[16], Ala[35], Ala[51]), the electrostatic nature of the mutation was opposite. In the case of 1csp, the electrostatic nature of the mutation was opposite in all four mutations (Leu[2], Ser[11], Phe[15], Ala[46]).

Comparison of the pattern of the residue differences ($\Delta E_i = E_i^{Th} - E_i^{Me}$) in the interaction energies between the mesophilic protein and its thermophilic counterpart for the natural cases (Figure 6.15) and the respective designed cases (Figure 6.16) clearly indicates that the stabilizing effects of the individual mutations are more pronounced in the case of designed thermophilic proteins compared to those of natural thermophilic proteins.

Comparison of the results indicates that there is good agreement between our designed protein and the natural thermophilic protein. However, exact reproduction of the natural thermophilic protein is not possible for the following reasons: (1) Natural thermophilic proteins evolve based on many evolutionary requirements, but our design is based on a single requirement of improved thermostability. (2) Our design aims to generate a protein with the highest thermostability with the same number of mutations, while the natural protein is made in an optimal way maintaining the other factors as well. (3) Natural design seems to take care of many other factors related to the stability of a protein, while we considered only the electrostatic and steric contributions. Thus it is not expected that the designed protein will be identical to the natural one. However, in our cases, the designed and the natural proteins were found to be reasonably close, thus validating our design method in general.

Tables 6.5 and 6.6 summarize the residue ΔE_i values for all the nonmatching residue pairs between the protein pair, considering electrostatic and van der Waals interactions. It can be seen that in several cases the most strongly interacting pairs are found to be "sequence-wise local," and hence they are not likely to be important for stabilizing or destabilizing the protein structure due to changes in residues.

6.5   DISCUSSION

The design approach presented here has several advantages over other methods where sequence comparison plays a major role. This design strategy is based on the actual 3D structure of the original protein of interest. Furthermore, it takes care of

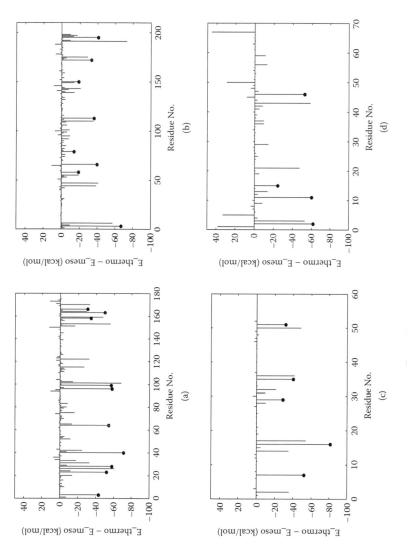

FIGURE 6.16 The pattern of the differences ($\Delta E_i = E_i^{Th} - E_i^{Me}$) in the residue interaction energies between the natural mesophilic protein and its designed thermophilic counterpart for (a) 1ino, (b) 1qnm, (c) 8rxn, and (d) 1csp. The mutated residues are highlighted by filled circles at the bottom of the respective bars. It is clear that the stabilization effects of the individual mutations are more pronounced here compared to those in the cases of natural thermophilic proteins.

realistic physical interactions between the individual residues at the atomic level using standard force fields and thus provides an opportunity to enhance protein stability through interresidue interactions in the "spatially local" region. Consideration of realistic side-chain conformers and force fields allows us to find suitable substitutes of side chains to improve its interactions with the rest of the protein.

Selected mutations of only the residues close to the protein surface largely ensure that the mutated protein will maintain a 3D structure close to the corresponding original protein.

In designing a thermophilic version of a functional protein, it is important not to mutate any of the residues involved in the active site, as alteration of the active site may have adverse effects on its activity because of changes in the local physico-chemical environment.

In order to enhance stability further, the procedure cycle can be repeated a couple of times using the modified protein to increase the number of favorable mutations.

The present procedure can even be applied to a region where no weakly interacting polar or charged residue is found. In such a case, a preselected residue of any type in the region of interest should be replaced with a suitable polar or charged residue following our methods, and this should be considered as a key residue. Subsequently, some other residue in its vicinity should be identified and replaced by a side chain of electrostatically complementary nature, as described in this chapter.

Depending on the amino acid sequence and composition, some proteins can be turned into thermophilic ones in a single cycle of the procedure described here. In principle, several consecutive cycles of the procedure should be able to turn any protein into a thermophilic one.

In this chapter, we have considered the electrostatic part of interresidue interactions as the most important aspect of the design strategy. In addition, we have taken into account the van der Waals interactions among the residues. Hydrophobic interaction is another factor that is believed to be important for protein stability, but we have not considered that factor in this chapter. Hydrophobic interactions can be included in such a strategy even though it is not yet very straightforward to do so.

Moreover, we have considered only the enthalpic part of the free energy differences in designing thermophilic proteins, as our objective is to increase the stability of the protein in a rather qualitative way. For a more accurate and quantitative description of stability, entropic contributions should also be captured, as entropic effects may produce additional favorable or unfavorable contributions to the structural stability of proteins.

Similarly, the solvation contribution can be incorporated through the solvation free energy calculation using the linear interaction energy (LIE) method.[66] This may also be useful in identifying the polar and charged amino acid residues that interact poorly with the rest of the protein.

MD simulations were done in vacuum since the purpose here was to compare the properties of the original protein and the designed mutated protein under the same thermal conditions. For a more realistic comparison, one needs to perform time-consuming MD simulations in explicit solvents.

So far we have examined stabilization of single protein chains. However, there are many examples where the actual functional forms of proteins are multimeric. In that case, in order to maintain the activity at high temperature, it is not enough to stabilize the individual monomers, it also requires stabilization of the quaternary structure of the enzyme by improving the interactions between the monomers in the interfaces. There is also an example where mutation in the interface has stabilized the quaternary complex and its activity.[67] Our method is also capable of designing the stabilization of such interfaces. We will demonstrate that elsewhere.

6.6 CONCLUSION

A structure-based computational method for designing thermophilic proteins has been described and validated by computational measures. The present method of designing a thermophilic protein starting with the 3D structure of a mesophilic protein has been demonstrated to be quite effective in the case studies considered here. The enhancement in stability of the designed protein has been demonstrated by computational measures. The method appears to be simple and fast, but robust, and applicable to any kind of protein. Consideration of the side-chain conformer libraries along with electrostatic and van der Waals interactions makes the approach more straightforward and accurate. Comparison with the properties of a set of known mesophilic–thermophilic protein pairs validates the approach. This method is particularly useful when an enzyme is needed to be a thermophilic one and the experimental 3D structure of the selected mesophilic enzyme is available.

ACKNOWLEDGMENTS

The authors would like to thank Nandita Sinha for critically reading the manuscript.

REFERENCES

1. de Champdoré, M., M. Staiano, M. Rossi, and S. D'Auria. 2007. Proteins from extremophiles as stable tools for advanced biotechnological applications of high social interest. *J R Soc Interface* 4: 183–191.
2. Jaenicke, R. 2000. Do ultrastable proteins from hyperthermophiles have high or low conformational rigidity? *Proc Natl Acad Sci USA* 97: 2926–64.
3. Szilagyi, A., and P. Zavodszky. 2000. Structural differences between mesophilic, moderately thermophilic and extremely thermophilic protein subunits: Results of a comparative survey. *Structure* 8: 493–504.
4. Fitter, J., and J. Heberle. 2000. Structural equilibrium fluctuations in mesophilic and thermophilic-amylase. *Biophys J* 79: 1629–36.
5. Vieille, C., and G. J. Zeikus. 2001. Hyperthermophilic enzymes: Sources, uses, and molecular mechanisms for thermostability. *Microbiol Mol Biol Rev* 65: 1-43.
6. Vogt, G., S. Woell, and P. Argos. 1997. Protein thermal stability, hydrogen bonds, and ion pairs. *J Mol Biol* 269: 631–43.
7. Spassov, V. Z., A. D. Karshikoff, and R. Ladenstein. 1995. The optimization of protein-solvent interactions: Thermostability and the role of hydrophobic and electrostatic interactions. *Protein Sci* 4: 1516–27.

8. Kumar, S., and R. Nussinov. 2001. How do thermophilic proteins deal with heat? *Cell Mol Life Sci* 58: 1216–33.

9. Missimer, J. H., M. O. Steinmetz, R. Baron, F. K. Winkler, R. A. Kammerer, X. Daura, and W. F. van Gunsteren. 2007. Configurational entropy elucidates the role of salt-bridge networks in protein thermostability. *Protein Sci* 16: 1349–59.

10. Perl, D., U. Mueller, U. Heinemann, and F. X. Schmid. 2000. Two exposed amino acid residues confer thermostability on a cold shock protein. *Nat Struct Biol* 7: 380–83.

11. Park, S., Y. Xu, X. F. Stowell, F. Gai, J. G. Saven, and E. T. Boder. 2006. Limitations of yeast surface display in engineering proteins of high thermostability. *Protein Eng Des Sel* 19: 211–17.

12. Razvi, A., and J. M. Scholtz. 2006. Lessons in stability from thermophilic proteins. *Protein Sci* 15: 1569–78.

13. Kannan, N., and S. Vishveshwara. 2000. Aromatic clusters: A determinant of thermal stability of thermophilic proteins. *Protein Eng* 13: 753–61.

14. Thomas, A. S., and A. H. Elcock. 2004. Molecular simulations suggest protein salt bridges are uniquely suited to life at high temperatures. *J Am Chem Soc* 126: 2208–14.

15. Gribenko, A. V., and G. I. Makhatadze. 2007. Role of the charge-charge interactions in defining stability and halophilicity of the CspB proteins. *J Mol Biol* 366: 842–56.

16. Grimsley, G. R., K. L. Shaw, L. R. Fee, R. W. Alston, B. M. Huyghues-Despointes, R. L. Thurlkill, J. M. Scholtz, and C. N. Pace. 1999. Increasing protein stability by altering long-range coulombic interactions. *Protein Sci* 8: 1843–49.

17. Perl, D., and F. X. Schmid. 2001. Electrostatic stabilization of a thermophilic cold shock protein. *J Mol Biol* 313: 343–57.

18. Dominy, B. N., H. Minoux, and C. L. Brooks, III. 2004. An electrostatic basis for the stability of thermophilic proteins. *Proteins* 5: 128–41.

19. Karshikoff, A., and R. Ladenstein. 1998. Proteins from thermophilic and mesophilic organisms essentially do not differ in packing. *Protein Eng* 11: 867–72.

20. Vogt, G., and P. Argos. 1997. Protein thermal stability: Hydrogen bonds or internal packing? *Fold Des* 2: S40–46.

21. Robinson, M. R., R. P. Eaton, D. M. Haaland, G. W. Koepp, E. V. Thomas, B. R. Stallard, and P. L. Robinson. 1992. Noninvasive glucose monitoring in diabetic patients: A preliminary evaluation. *Clin Chem* 38: 1618–22.

22. Turner, P., G. Mamo, and E. N. Karlsson. 2007. Potential and utilization of thermophiles and thermostable enzymes in biorefining. *Microbial Cell Factories* 6: 9. doi:10.1186/1475-2859-6-9.

23. Chien, A., D. B. Edgar, and J. M. Trela. 1976. Deoxyribonucleic acid polymerase from the extreme thermophile *Thermus aquaticus*. *J Bacteriol* 127: 1550–57.

24. Kaledin, A. S., A. G. Sliusarenko, and S. I. Gorodetskii. 1980. Isolation and properties of DNA polymerase from extreme thermophylic bacteria *Thermus aquaticus* YT-1. *Biokhimiya* 45: 644–51.

25. Lehmann, M., L. Pasamontes, S. F. Lassen, and M. Wyss. 2000. The consensus concept for thermostability engineering of proteins. *Biochim Biophys Acta* 1543: 408–15.

26. Lehmann, M., C. Loch, A. Middendorf, D. Studer, S. F. Lassen, L. Pasamontes, A. P. G. M. van Loon, and M. Wyss. 2002. The consensus concept for thermostability engineering of proteins: Further proof of concept. *Protein Eng* 15: 403–11.

27. Korkegian, A., M. E. Black, D. Baker, and B. L. Stoddard. 2005. Computational thermostabilization of an enzyme. *Science* 308: 857–60.

28. Zollars, E. S., S. A. Marshall, and S. L. Mayo. 2006. Simple electrostatic model improves designed protein sequences. *Protein Sci* 15: 2014–18.

29. Dantas, G., B. Kuhlman, D. Callender, M. Wong, and D. Baker. 2003. A large scale test of computational protein design: Folding and stability of nine completely redesigned globular proteins. *J Mol Biol* 332: 449–60.

30. Eijsinka, V. G. H., A. Bjørk, S. Gåseidnes, R. Sirevåg, B. Synstad, B. van den Burg, and G. Vriend. 2004. Rational engineering of enzyme stability. *J Biotechnol* 113: 105–20.
31. Eijsinka, V. G. H., S. Gåseidnes, T. V. Borchert, and B. van den Burg. 2005. Directed evolution of enzyme stability. *Biomol Eng* 22: 21–30.
32. Montanucci, L., P. Fariselli, P. L. Martelli, and R. Casadio. 2008. Predicting protein thermostability changes from sequence upon multiple mutations. *Bioinformatics* 24: i190–95.
33. Lehmann, M., and M. Wyss. 2001. Engineering proteins for thermostability: The use of sequence alignments versus rational design and directed evolution. *Curr Opin Biotechnol* 12: 371–75.
34. Jäckel, C., J. D. Bloom, P. Kast, F. H. Arnold, and D. Hilvert. 2010. Consensus protein design without phylogenetic bias. *J Mol Biol* 399: 541–46.
35. Chan, C.-H., H.-K. Liang, N.-W. Hsiao, M.-T. Ko, P.-C. Lyu, and J.-K. Hwang. 2004. Relationship between local structural entropy and protein thermostability. *Proteins* 57: 684–91.
36. Bannen, R. M., V. Suresh, G. N. Phillips, Jr., S. J. Wright, and J. C. Mitchell. 2008. Optimal design of thermally stable proteins. *Bioinformatics* 24: 2339–43.
37. Bae, E., R. M. Bannen, and G. N. Phillips, Jr. 2008. Bioinformatic method for protein thermal stabilization by structural entropy optimization. *Proc Natl Acad Sci USA* 105: 9594–97.
38. Shah, P. S., G. K. Hom, S. A. Ross, J. K. Lassila, K. A. Crowhurst, and S. L. Mayo. 2007. Full-sequence computational design and solution structure of a thermostable protein variant. *J Mol Biol* 372: 1–6.
39. Berezovsky, I. N., K. B. Zeldovich, and E. I. Shakhnovich. 2007. Positive and negative design in stability and thermal adaptation of natural proteins. *PLoS Comput Biol* 3: e52. doi:10.1371/journal.pcbi.0030052.
40. Kuhlman, B., G. Dantas, G. C. Ireton, G. Varani, B. L. Stoddard, and D. Baker. 2003. Design of a novel globular protein fold with atomic-level accuracy. *Science* 302: 1364–68.
41. Butterfoss, G. L., and B. Kuhlman. 2006. Computer-based design of novel protein structures. *Annu Rev Biophys Biomol Struct* 35: 49–65.
42. Dahiyat, B. I., and S. L. Mayo. 1997. *De novo* protein design: Fully automated sequence selection. *Science* 278: 82–87.
43. Basu, S., and S. Sen. 2009. Turning a mesophilic protein into a thermophilic one: A computational approach based on 3D structural features. *J Chem Inform Model* 49: 1741–50.
44. Xiao, L., and B. Honig. 1999. Electrostatic contributions to the stability of hyperthermophilic proteins. *J Mol Biol* 289: 1435–44.
45. Karshikoff, A., and R. Ladenstein. 2001. Ion pairs and the thermotolerance of proteins from hyperthermophiles: A "traffic rule" for hot roads. *Trends Biochem Sci* 26: 550–56.
46. Elcock, A. H. 1998. The stability of salt bridges at high temperatures: Implications for hyperthermophilic proteins. *J Mol Biol* 284: 489–502.
47. Danciulescu, C., R. Ladenstein, and L. Nilsson. 2007. Dynamic arrangement of ion pairs and individual contributions to the thermal stability of the cofactor-binding domain of glutamate dehydrogenase from *Thermotoga maritime*. *Biochemistry* 46: 8537–49.
48. Tanner, J. J., R. M. Hecht, and K. L. Krause. 1996. Determinants of enzyme thermostability observed in the molecular structure of *Thermus aquaticus* D-glyceraldehyde-3-phosphate dehydrogenase at 2.5 Å resolution. *Biochemistry* 35: 2597–609.
49. Robb, F. T., and D. S. Clark. 1999. Adaptation of proteins from hyperthermophiles to high pressure and high temperature. *J Mol Microbiol Biotechnol* 1: 101–5.

50. Loladze, V. V., B. Ibarra-Olero, J. M. Sanchez-Ruiz, and G. I. Makhatadze. 1999. Engineering a thermostable protein via optimization of charge-charge interactions on protein surface. *Biochemistry* 38: 16419–23.

51. Spector, S., M. Wang, S. A. Carp, J. Robblee, Z. S. Hendsch, R. Fairman, B. Tidor, and D. P. Raleigh. 2000. Rational modification of protein stability by mutation of charged surface residues. *Biochemistry* 39: 872–79.

52. Strickler, S. S., A. V. Gribenko, T. R. Keiffer, J. Tomlinson, T. Reihle, V. V. Loladze, and G. I. Makhatadze. 2006. Protein stability and surface electrostatics: A charged relationship. *Biochemistry* 45: 2761–66.

53. Strub, C., C. Alies, A. Lougarre, C. Ladurantie, J. Czaplicki, and D. Fournier. 2004. Mutation of exposed hydrophobic amino acids to arginine to increase protein stability. *BMC Biochem* 5: 9. doi:10.1186/1471-2091-5-9.

54. Makhatadze, G. I., V. V. Loladze, A. V. Gribenko, and M. M. Lopez. 2004. Mechanism of thermostabilization in a designed cold shock protein with optimized surface electrostatic interactions. *J Mol Biol* 336: 929–42.

55. Tanaka, T., M. Sawano, K. Ogasahara, Y. Sakaguchi, B. Bagautdinov, E. Katoh, C. Kuroishi, S. Shinkai, S. Yokoyama, and K. Yutani. 2006. Hyper-thermostability of CutA1 protein, with a denaturation temperature of nearly 150 degrees C. *FEBS Lett* 580: 4224–30.

56. Brooks, B. R., R. E. Bruccoleri, B. D. Olafson, D. J. States, S. Swaminathan, and M. Karplus. 1983. CHARMM: A program for macromolecular energy, minimization, and dynamics calculations. *J Comput Chem* 4: 187–217.

57. MacKerell, A. D., Jr., D. Bashford, M. Bellott, R. L. Dunbrack, Jr., J. D. Evanseck, M. J. Field, S. Fischer, J. Gao, H. Guo, S. Ha, D. Joseph-McCarthy, L. Kuchnir, K. Kuczera, F. T. K. Lau, C. Mattos, S. Michnick, T. Ngo, D. T. Nguyen, B. Prodhom, W. E. Reiher III, B. Roux, M. Schlenkrich, J. C. Smith, R. Stote, J. Straub, M. Watanabe, J. Wiórkiewicz-Kuczera, D. Yin, and M. Karplus. 1998. All-atom empirical potential for molecular modeling and dynamics studies of proteins. *J Phys Chem B* 102: 3586–616.

58. Sen, S., and L. Nilsson. 1999. Structure, interactions, dynamics and solvent effects of the DNA-EcoRI complex in aqueous solutions from molecular dynamics simulation. *Biophys J* 77: 1782–800.

59. Sen, S. 2002. Exploring structure and energetics of a helix forming oligomer by MM and MD simulation methods: Dynamics of water in a hydrophobic nanotube. *J Phys Chem B* 106: 11343–50.

60. Roy, S., and S. Sen. 2005. Homology modeling based solution structure of Hoxc8-DNA complex: Role of context bases outside TAAT stretch. *J Biomol Struct Dynam* 22: 707–18.

61. Roy, S., and S. Sen. 2006. Exploring the potential of complex formation between a mutant DNA and the wild-type protein counterpart: A MM and MD simulation approach. *J Mol Graph Model* 25: 158–68.

62. Hockney, R. W. 1970. The potential calculation and some applications. *Methods Comput Phys* 9: 136–211.

63. Van Gunsteren, W. F., and H. J. C. Berendsen. 1977. Algorithms for macromolecular dynamics and constraint dynamics. *Mol Phys* 23: 1311–27.

64. Ryckaert, J. P., G. Ciccotti, and H. J. C. Berendsen. 1877. Numerical integration of the Cartesian equations of motion of a system with constraints: Molecular dynamics of n-alcanes. *J Comput Phys* 23: 327–41.

65. Steinbach, P. J., and B. R. Brooks. 1994. New spherical cutoff methods for long range forces in macromolecular simulation. *J Comp Chem* 15: 667–683.

66. Bren, U., V. Martinek, and J. Florian. 2006. Free energy simulations of uncatalyzed DNA replication fidelity: Structure and stability of T·G and dTTP·G terminal DNA mismatches flanked by a single dangling nucleotide. *J Phys Chem B* 110: 10557–66.
67. Kaneko, H., H. Minagawa, and J. Shimada. 2005. Rational design of thermostable lactate oxidase by analyzing quaternary structure and prevention of deamidation. *Biotechnology Lett* 27: 1777–84.

Index

A

Acidothermus cellulolyticus, 112
Adenylate cyclase, 110–111
Aeromonas sobria, 81
AkP protease, 84, 89, 91, 94
alpha-LP, 76
Amylase, 31, 118
Aqualysin I (AQUI), 71
    C-domain, sequence homology of B domain
        of Tk-SP and, 82
    calcium-binding sites, 87, 88
    disulfides, 93–94
    mutations, 92
    N-terminal regions, 93
    processing, 72
    prodomain of, 72
    proline residues, 92
    salt bridge, 91
    structure, 90, 91–92
        stabilizing effect of sodium ions on, 89,
          90
    VPR *vs.,* 91, 93
AQUI; *See* Aqualysin I
Aquifex sp., 2
Aquifex aerolicus, 9
Archaea, 2, 11
    electron-transferring protein, 58
    mesophilic *vs.* thermophilic, 30

B

B-FITTER software, 108
Bacillus sp., 80
Bacillus caldolyticus, 9, 34
Bacillus cereus, 54
Bacillus sphaericus, 80
Bacillus subtilis, 9
    lipase from mesophilic, 40
Bacillus TA41, 75, 80
Bacillus WF146, 80

C

Calcium binding, 69, 80–89
Calorimetry, 29
CATH Data Bank, 69
CelC, 109
    B-factors, 109

E. coli expression, 109
    enzymatic activity, reduction in, 109
    gene, 108, 109
    parenteral activity, reduction of, 110
    thermal stability in, 113–114
    three crystallographic structures, 109
Cellulases, 111–113
CHARMM software, 122
Circular dichroism spectroscopy, 3, 82, 89
Clostridium cellulovorans, 112
Clostridium pasteurianum, 30, 58
Clostridium thermocellum, 109
CNA; *See* Constraint network analysis
Cold shock protein, 34
    folding kinetics, 31
    mesophilic–thermophilic pair, 139
    thermodynamic stability, 9
Computer simulation, 22–25
Conformational flexibility; *See also* Flexibility
    enzymatic activity and, 30
    protein stability and, 59
Consensus method, 119
Constraint network analysis (CNA), 49–50
    applications, 54–60
        rigidity theory and, 58–60
    thermal unfolding simulation based on,
        50 53
Cys99Ser, 94
Cys194Ser, 94

D

DCM; *See* Distance constraint model
Dielectric constant, 37–38
Differential scanning calorimetry (DSC), 3, 77,
    92, 112
Directed evolution, 107–108
Distance constraint model (DCM), 59–60
Disulfide(s), 93–94
    bonds, 69
        AkP protease, 94
        function of, 92
        location of, 92
DSC; *See* Differential scanning calorimetry

E

Electrostatics, 33–38
    flexibility and activity, 38

155

Printed and bound by CPI Group (UK) Ltd, Croydon, CR0 4YY

18/10/2024

01776271-0002